I0888976

Saving the Pryor Mountain Mustang

Saving the Pryor Mountain Mustang

A Legacy of Local and Federal Cooperation

Christine Reed

UNIVERSITY OF NEVADA PRESS *Reno & Las Vegas*

University of Nevada Press, Reno, Nevada 89557 USA
Copyright © 2015 by University of Nevada Press
All rights reserved
Manufactured in the United States of America
Design by Kathleen Szawiola

Library of Congress Cataloging-in-Publication Data
Reed, Christine M. (Christine Mary).
Saving the Pryor Mountain mustang : a legacy of local and federal cooperation / Christine M. Reed.
pages cm
Includes bibliographical references and index.
ISBN 978-0-87417-966-8 (cloth : alk. paper) —
ISBN 978-0-87417-967-5 (e-book)
1. Wild horses—Pryor Mountain Wild Horse Range (Mont. and Wyo.)—Management. 2. Mustang—Pryor Mountain Wild Horse Range (Mont. and Wyo.)—Management. 3. Pryor Mountain Wild Horse Range (Mont. and Wyo.)—History—Sources. 4. Wildlife conservation—Government policy—United States. I. Title.
SF360.3.U6R44 2014
636.1'3097874—dc23 2014032928

The paper used in this book meets the requirements of American National Standard for Information Sciences—Permanence of Paper for Printed Library Materials, ANSI/NISO Z39.48-1992 (R2002). Binding materials were selected for strength and durability.

FIRST PRINTING
23 22 21 20 19 18 17 16 15
5 4 3 2 1

This book is dedicated to the memory of

REVEREND FLOYD SCHWIEGER

October 3, 1920–July 20, 2005

Contents

Preface

In the summer of 2009, I attended my first wild horse photography workshop. I had just finished reading *Among Wild Horses,* a photographic journey with family bands and bachelor groups on the Pryor Mountain Wild Horse Range (PMWHR). When I looked up the author's name on the Internet, I was happy to discover that Lynne Pomeranz also offered photography workshops. As an aspiring photographer, I was thrilled at the prospect. As a professor of public administration, however, I could not resist the impulse to download the government documents related to the PMWHR from the Bureau of Land Management's (BLM) website. Before the workshop began, we met at the Pryor Mountain Wild Mustang Center (PMWMC) in Lovell, Wyoming, with Jared Bybee, BLM wild horse specialist, who provided us with a brief history of the range and its horses.

As I started asking him questions about the *Herd Management Area Plan* (*HMAP*), a document whose details I had absorbed before driving to the workshop, I noticed that the other members of the workshop were staring at me. So I asked the obvious question: was I the only one who had downloaded and read this plan? Their reply (in unison) was yes. I realized at that moment that it would be impossible for me to separate my photography from my academic life. After the workshop I presented some of my photographs at a gallery exhibit, but I also began writing, presenting, and publishing papers comparing government management of wild horses in the United States with the Netherlands, where descendants of extinct wild horses are being released back into public nature reserves, in contrast with the US policy of rounding them up. I returned to Lovell and to the wild horse range every year thereafter for research and photography.

In 2011 I presented a paper to the Western Historical Association (WHA) annual conference in Oakland, California, about the local efforts of Lovell, Wyoming, citizens in the 1960s to save a unique herd of wild horses on the PMWHR. The public management of wild horses, not only in the United States but also in Europe, is a hot-button issue, raising a host of ethical and policy questions. Exchanges between wild horse activists and government agencies are often contentious. The most common form of interaction, it seems, is litigation. The case of Lovell, Wyoming, is unusual, however, in that local wild horse advocates chose to work with, rather than against, the BLM. The earliest Lovell advocates contributed to the formation of the PMWHR, the first public wild horse refuge in the United States. They established the Colonial Spanish lineage of horses in the Pryor wild herd, linking them to the battle mounts of the conquistadores.

I met my acquisition editor, Matt Becker, at the 2011 WHA conference. He became interested in my proposal for a book, because the story of the Pryor Mountain Wild Horse Range is so unique, pointing to possibilities for federal-local cooperation in the public management of America's wild horses. As is the case with any book project, there are many other individuals who supported this project and whom I wish to thank. I am grateful to the Pryor Mountain Wild Mustang Center for welcoming me on a daily basis as I hunkered down in the back of the building, going through mountains of archival materials. I am also very grateful to those who granted me interviews and whose identities I must keep confidential to honor my assurances to them. As the saying goes, "You know who you are."

I owe a special thank-you to four individuals who "fact-checked" the entire manuscript. The first three are John Nickle, the longtime center board president and "institutional memory"; Matt Dillon, environmental scientist and the center's first director, who has an unparalleled grasp of the lineage of every horse—past and present—on the range; and Nancy Cerroni, board member and educational administrator, who carries on a tradition started by Velma "Wild Horse Annie" Johnston of engaging schoolchildren across the country in the cause of wild horses.

My fourth "fact-checker" is Jared Bybee, the rangeland management specialist and Montana-Dakotas state wild horse specialist, who leads the way in twenty-first-century management of wild horses. His contribution provided an important BLM perspective on the events of the 1960s, particularly during the years leading up to the formation of the PMWHR. He also made available a complete administrative record of the period from 1968 to 2009. These four individuals provided me with valuable feedback.

I would be remiss if I failed to mention Bill and Kathy Johnsey, who allowed me to stay in their camper on their farm outside of Lovell. The camper rested on the shore of Lake Lovell and, besides providing me affordable living quarters for two summers, offered me thrilling views right outside my door of osprey, white pelicans, great blue herons, and many species of ducks and geese. I also met their friends and neighbors during the summers I spent there, making those months a lot less lonely than they otherwise might have been. I also became a regular at the Horseshoe Bend motel during five summer wild horse photography workshops, and I am grateful for the good care Joe and Jo took of me.

Of course, I have had great support from the School of Public Administration at the University of Nebraska, Omaha, throughout this process, too, including a semester of professional development leave to complete the project. As anyone who has embarked on the journey of writing a book can attest, people who can troubleshoot their offices' temperamental copiers and can decode the mysteries of word processing are the midwives to authors who labor to get out their final creations. I want to thank Dean John Bartle, school director Ethel Williams, staff assistant Rhonda Sheibal-Carver, and doctoral assistant Anthony Campbell.

Working with the University of Nevada Press has been a delight. The professionals who "walked me" through the intricate process of publication were experts in their crafts, and they helped to make an otherwise daunting project a pleasant experience. I want to extend a special thanks to senior acquisitions editor Matt Becker, design and production manager Kathleen Szawiola, and Annette Wenda, my copy editor. The publishing staff have ensured that I can be proud of this book project.

Finally, book authors thank—or even dedicate their work to—their families for a very good reason. I suspect it dawns on them when their manuscripts are complete how immersed they became in their projects and how their spouses, partners, and children were encouraging them, if only by withholding comments about their significant others being in absentia until the final product was finally in press. It is true what they say, that no book is ever possible without the contributions of many professional and personal supporters. Like authors before me, I thank those who supported me.

Introduction

This fifty-year history of consultation between Lovell advocates and the BLM is a single case study with multiple lessons to be learned about the public management of wild horses. The extent to which this process of consultation is responsible for twenty-first-century practices being implemented on the Pryor Mountain Wild Horse Range is, of course, hypothetical. It is impossible to unravel the causal effects of human relationships on management outcomes over such a lengthy time frame. Nevertheless, archival documents and interviews support the main thesis of this book—that local consultation with BLM field office staff added value to decisions made under the authority of the Wild Free Roaming Horses and Burros Act (WFRHBA). The main focus of this book is the sustained effort by Lovell advocates to incorporate their goal of preserving the Pryor Mountain Mustang into BLM decision making at the Billings, Montana, field office, the agency with the primary responsibility for the Pryor wild herd.

By 2008 the BLM had formally recognized the goals it held in common with the Pryor Mountain Wild Mustang Center in a Memorandum of Understanding calling on Lovell advocates to continue monitoring and recording information about the Pryor horses, reaching out to the public with information about this unique herd, and assisting the BLM in other ways. Center board members and staff have contributed data from its genealogy project as part of an effort to develop criteria for removing horses from the range. Today, the Billings Field Office is able to determine the lineage of each horse on the range, ensuring that periodic removals preserve all bloodlines now represented in the herd. This consultation, it should be noted, has been on a volunteer basis.

Use of the term *consultation* is deliberate in order to distinguish the process used by Lovell advocates from the broader concept of collaboration. In the field of public administration, the process of collaboration, particularly in the area of environmental policy, is associated with the trend away from top-down government regulation to horizontal *governance* networks. This trend started in the 1990s with the decentralization of government power from federal to state and local levels, as well as the privatization of formerly public services. While consultation with a nonprofit entity by a government agency is a fairly traditional form of citizen engagement, the process of collaboration often involves shared control over decision making. Federal laws and regulations governing the public management of wild horses require opportunities for public comment; however, devolution of authority is limited, and privatization is restricted to contracts with long-term holding facilities and pastures.

Authorizing legislation leaves BLM field offices with little discretion for deviating from rules governing the removal of excess horses. Despite this legal framework, many wild horse advocates insist that the BLM is authorized to suspend or eliminate wild horse roundups. In this national climate of alienation and suspicion, the fifty-year history of Lovell advocates and the BLM exemplifies the potential for building trust and working together to develop plans for nurturing the health of the Pryor Mountain Mustangs and the land on which they depend for their survival. Wyoming culture is fiercely independent. Nevertheless, its people believe that differences between neighbors can be worked out through conversation. Lovell advocates have never been shy about seeking federal protection for the wild herd and the expansion of the range, but they value consultation over confrontation.

One unintended consequence of this local culture was that Lovell became "ground zero" for a time as a result of the "Cloud" documentaries filmed on the PMWHR and aired on the PBS *Nature* series. The devotion of national viewers to this palomino stallion sparked protests during a large-scale removal in 2009, and Lovell advocates as well as the BLM became targets of public outrage because the center provided volunteer assistance during the roundup. On-site protests, critical messages sent to the center, and face-to-face confrontations between activists and the more visible leaders in Lovell left residents with an abiding distrust of outside activists. It took several summers visiting Lovell to establish my credibility with those who responded to my request for interviews.

My approach to contacting key actors was dictated by the Human Subjects Review Board at my university: a letter or e-mail outlining my research project, an invitation (no "cold calls") to participate by face-to-face interview, a field study guide with the questions I wanted to ask, and a promise of confidentiality. I interviewed past and present members of the Pryor Mountain Wild Mustang Center, the Pryor Mountain Mustang Breeders Association (PMMBA), news media, and officials from the Billings Field Office of the BLM. The center's archives held another major source of data: numerous letters, minutes from meetings, and articles dating back to the 1960s. What I found was that people, while initially reserved, were open to conversation. My approach was to examine the issues from the multiple standpoints of those who are involved at the ground level, trying to understand their own lived experiences.

Contextualizing history in this way will, I am fairly certain, irritate some people who are impatient with the slow progress of reform. Other books about wild horses on the market are more critical about these matters. A more critical stance, however, would mean discounting local understandings as mere subjective perceptions. Critical theorists would say these are a gullible people, unaware of how powerful financial interests control decisions about multiple uses of public lands. Local consultation by Lovell advocates would, according to this more critical stance, be either naive or complicit in BLM efforts to rid public rangelands of wild horses. My own philosophy about acquiring and validating knowledge is that there is no "god's-eye view" allowing one to claim objectivity in the critical analysis of any social phenomenon.

I have tried to document the process of consultation with the BLM used by one community over a nearly fifty-year period. Whether these efforts could be replicated elsewhere or not, the ideas generated by Lovell advocates in consultation with the Billings Field Office are, in many ways, twenty-first-century approaches: managing for improved rangeland conditions as a way to allow more wild horses in a herd management area, on-the-range population measures using temporary fertility control, and attention to bloodlines as a way of conserving genetic diversity. It is worth pointing out that many critics who once opposed these approaches because they smacked of excessive human manipulation are now supportive of such measures. The Cloud Foundation (TCF), for example, is now in favor of combining PZP (porcine zona pellucida) with adaptive management: experimenting and adjusting on-the-range actions based on monitoring their effects on horse populations. Former critics may therefore find common ground with Lovell advocates, adding value to the public management of wild horses.

In documenting this fifty-year history, I have organized the material chronologically. The first chapter explores the origin of the Pryor wild horses, first observed by homesteaders in the forests north of what is now the town of Lovell, Wyoming. In order to understand why the BLM was under pressure to remove these wild horses, it is necessary to explain federal and state livestock rules in effect from the 1930s to the 1960s. The first chapter sets the stage for more recent history, recounting how a family of cattle ranchers refused to remove the Pryor wild herd from their BLM livestock allotment and how a small group of advocates representing the Lovell Chamber of Commerce came to their aid. The 1968 Secretarial Order of Stewart Udall created the first public wild horse range, protecting the wild herd but leaving many decisions to the BLM field office responsible for managing the new PMWHR.

The second chapter covers the period from the early 1970s to the 1990s. Passage of the Wild Free Roaming Horses and Burros Act in 1971 protected wild horses where they were "presently found," allowing for expansion of the original range boundaries to include several thousand acres in the adjacent national forest. The range also included part of the Bighorn Canyon National Recreation Area, managed by the National Park Service (NPS). A tripartite legal arrangement reflecting diverse agency mandates resulted in years of consultation with Lovell advocates over such questions as how to balance the numbers of wild horses and bighorn sheep in the national park section, whether wild horses had in fact roamed in parts of the national forest that the US Forest Service (USFS) wanted to close for the protection of potentially endangered plant species, and what was an appropriate or sustainable number of wild horses on semiarid and subalpine rangelands under BLM jurisdiction.

During the early 1990s Lovell advocates formed two influential organizations, the Pryor Mountain Wild Horse Association (PMWHA) and the Pryor Mountain Mustang Breeders Association. Even though members saw early on that the wild herd had the characteristics of the Colonial Spanish horse, their arguments for preserving this unique herd were vindicated when they were able to use new blood-typing techniques to confirm that horses in the wild herd had genetic markers pointing to a lineage dating back to the battle horses of the conquistadores. They consulted with the Billings Field Office of the BLM, recommending which horses to keep on the range based on their physical characteristics or phenotype. They showcased their adopted Pryor horses in local parades and regional horse shows. Interest in preserving the Pryor Mountain Mustang peaked during this period, as did concern about the consequences of the BLM introducing wild horses from other herd areas in Wyoming and Oregon.

The third chapter builds on earlier themes, focusing on the "barn raising" of the Pryor Mountain Wild Mustang Center with contributions of money and human labor from area businesses, foundations, local government, and individual citizens. A cadre of volunteers raised funds, planned the design, formulated a detailed action plan, and joined the board of directors. In many ways they owed their inspiration to the Reverend Floyd Schwieger, whose leadership on behalf of the Pryor wild horses dated back to the 1960s and whose passion for preserving the Pryor Mountain Mustang led him to spend countless hours photographing, describing, and naming each wild horse. His death in 2005 made it all too clear how a small community-based nonprofit depends on the efforts of a single individual. After Schwieger's death, former town manager, council member, mayor, owner of a local cable company, and one of the original Lovell advocates John Nickle stepped up to lead the center's board of directors.

The theme of consultation runs throughout each chapter, along with the issues and challenges that emerged at different points in the history of the Pryor Mountain Wild Horse Range. In chapter 4 the center's philosophy of working with, not against, the BLM and a large-scale removal created the conditions for a "perfect storm" that blew through Lovell in 2009. Protests from the Cloud Foundation and other wild horse organizations created tension in the community, while direct criticism by outside groups of the center's first director polarized the situation even further. Critics charged that Matt Dillon, the center director who had assumed the role once played by the Reverend Floyd Schwieger, was working for the BLM, insinuating that the center was complicit in what they considered a violation of the wild horse law. What he was actually doing was identifying each horse as it came through the chute into the corrals, in order to make sure that family bands stayed together.

While the center's board and director were unhappy with certain aspects of the 2009 operation, they were faithful to their philosophy of continuing to make what difference they could as events unfolded. They continued to work with the Billings Field Office to implement the new *Herd Management Area Plan,* a major revision of previous documents that had the potential to improve conditions on the range and eliminate the need for any future helicopter roundups. For the next several years, Matt Dillon developed a genealogy project and posted a series of blog entries to educate the public about the scientific basis for wild horse management. As this period came to a close, the conditions for a perfect storm had dissipated; however, dark clouds continued to loom on the horizon. The center's board remained optimistic about the future, but some members foresaw a need to change the way they interacted with Lovell citizens and outside wild horse activists.

The fifth and final chapter speculates about issues and trends leading to the future. One big question hanging in the air is the future of the Pryor Mountain Mustang as a distinct strain of the Colonial Spanish horse. The BLM, with help from the center's genealogy project, has the capacity to continue managing for the representation of all existing bloodlines and colors on the range. Only a few Pryor Mountain Mustang breeders remain, however, located far from Lovell, in Nebraska and Massachusetts, as well as in Europe, where the ancestors of the Pryor horses originated. These individuals represent a new generation of Pryor breeders. The next chapter in that story remains to be written. Preservation of the Pryor Mountain Mustang was the singular goal of the early Lovell advocates; however, even the staunchest supporters of the wild herd are today as concerned about conserving genetic diversity and the overall health of the wild herd as they are in preserving the iconic primitive Spanish type of wild horse.

The future of consultation between Lovell advocates and the BLM is uncertain and depends on whether they can muster the resources, both human and financial, to continue monitoring the wild herd. Center board members are both optimistic and dedicated, but, like many in their shoes, they may find that survival depends on a new management style based on horizontal networks with other individuals and nonprofits. The challenge is to maintain its distinctive approach to wild horse management, while identifying organizational resources within and outside Lovell that will support (without duplicating) its mission. This challenge will be even harder for the center, which has succeeded until now by insulating itself from the drama of national politics that seems to follow controversy about the public management of wild horses. This fifty-year narrative is still in motion, its finale yet to be written.

Pryor Mountain Wild Horse Range Boundary: Secretarial Order

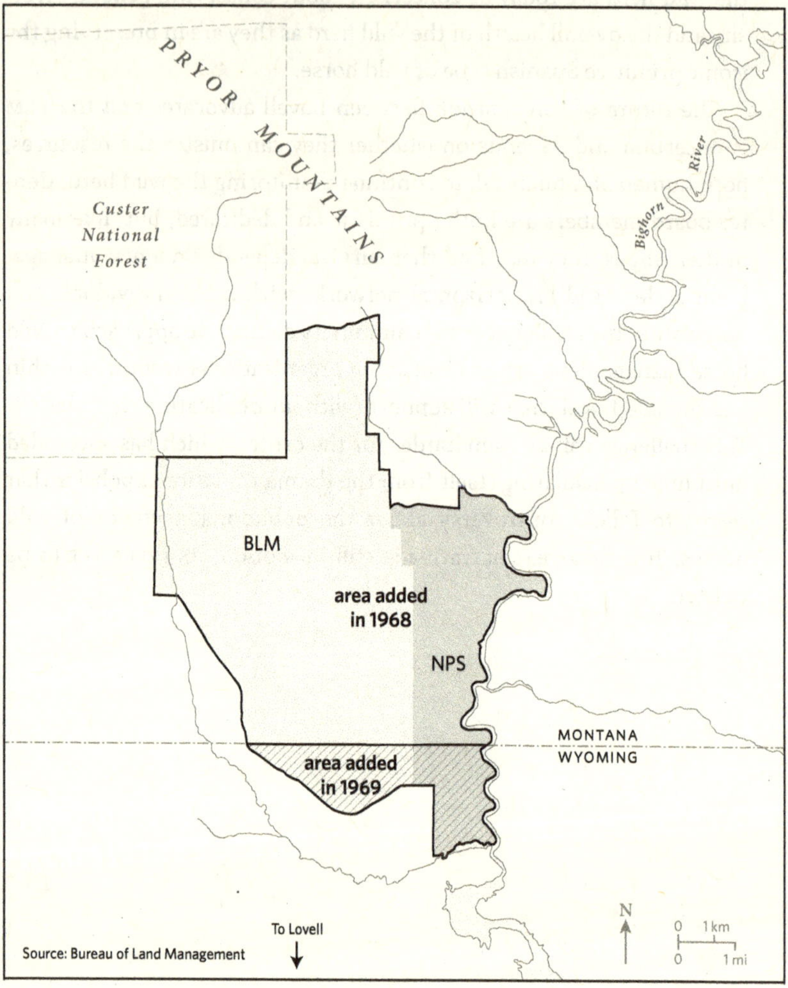

Pryor Mountain Wild Horse Range Boundary: Expansion After 1971

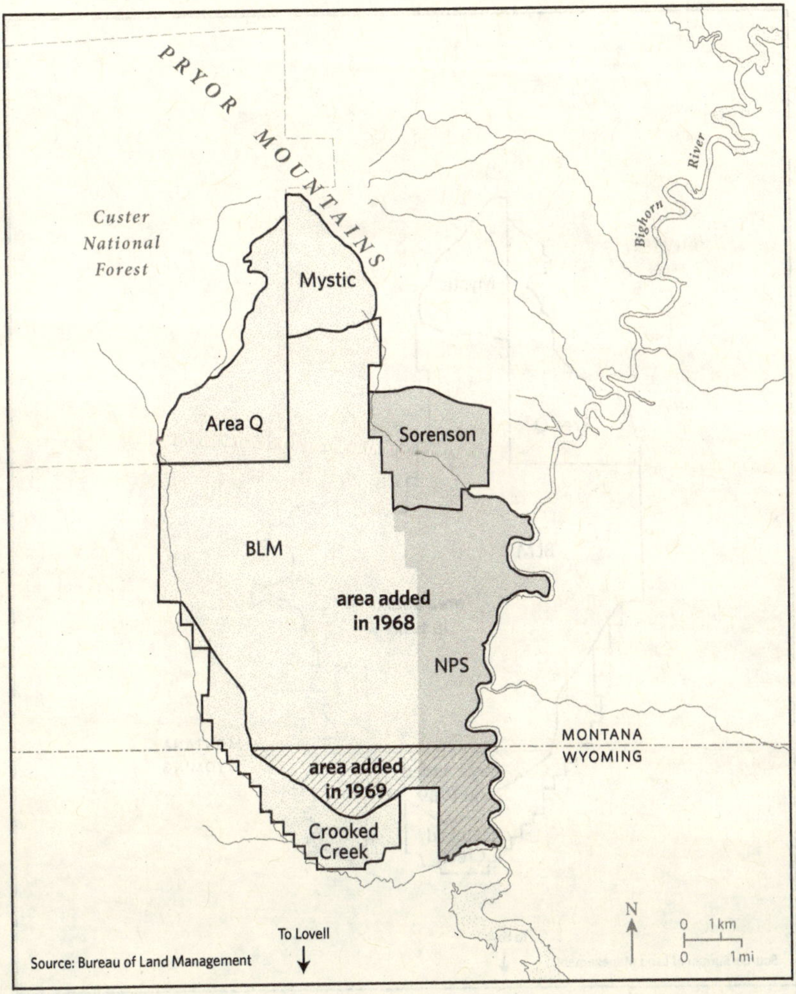

Pryor Mountain Wild Horse Range Boundary: Present Day

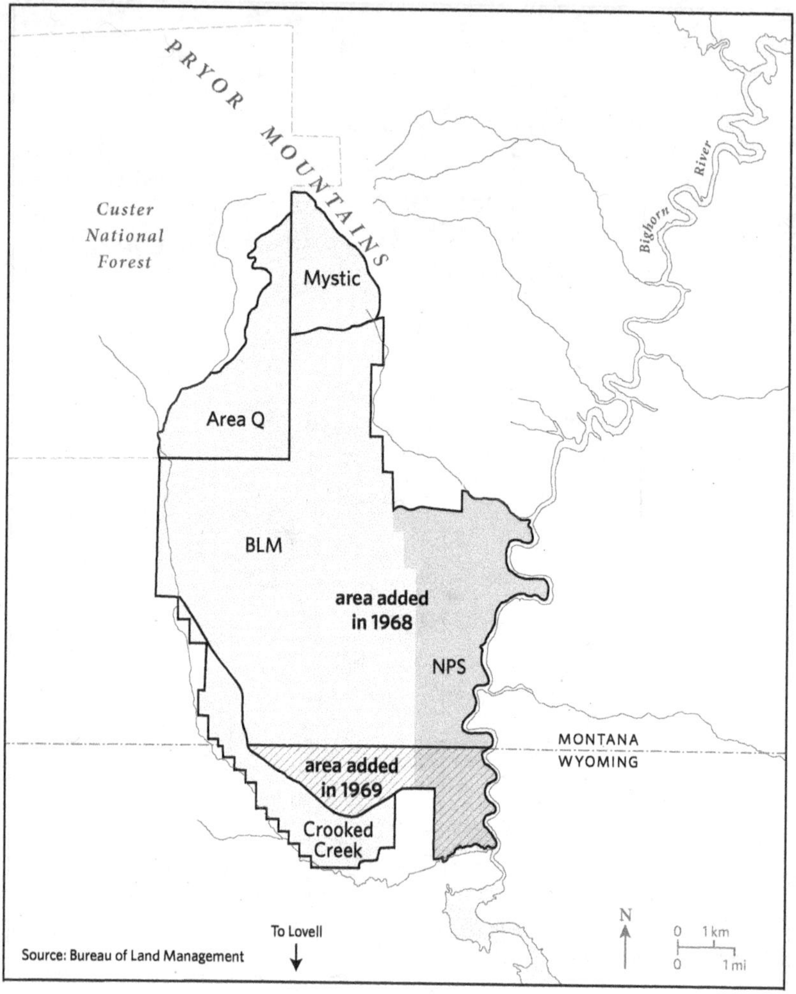

Abbreviations

AML	Appropriate Management Level
BLM	Bureau of Land Management
DRMP	Draft Resource Management Plan
EA	Environmental Assessment
EIS	Environmental Impact Statement
HMAP	*Herd Management Area Plan*
HSUS	Humane Society of the United States
ISPMB	International Society for the Protection of Mustangs and Burros
NPS	National Park Service
NRA	National Recreation Area
PMMA	Pryor Mountain Mustang Association
PMMBA	Pryor Mountain Mustang Breeders Association
PMWHA	Pryor Mountain Wild Horse Association
PMWHC	Pryor Mountain Wild Horse Center
PMWHR	Pryor Mountain Wild Horse Range
PMWMC	Pryor Mountain Wild Mustang Center
ROAM	Restore Our American Mustangs Act
TCF	The Cloud Foundation
USFS	U.S. Forest Service
WFRHBA	Wild Free Roaming Horses and Burros Act
WHOA	Wild Horse Organized Assistance

Saving the Pryor Mountain Mustang

Chapter One

Trespass Horses

Advocating for the Pryor Mountain Mustang

In a 1997 article for the *Casper Star Tribune*, staff writer Tom Mast penned the following poetic lines:

> In the outland there lies a city of roses beside a kingdom of mustangs. When winter melts away, residents of Lovell turn the soil to encourage a profusion of roses that will grace summer walkways. Nearby, gray and buckskin foals, the offspring of wild horses, test their spindly legs on ancient tepee rings scattered in mountain pastures.
>
> There is a circle to growing roses and wild horses. Mustangs and roses are intertwined with people's lives, on official welcome signs, over shop doors, cheered by middle school fans and high school revelers at Mustang Days, or in an artist's hand which forms the likeness of a free-roaming stallion, glistening in the sun. The lines flow together, continuous and complete.
>
> But once upon a time, the circle was nearly broken. The horses were imperiled; when in their hour of greatest need, the people forged a shield.[1]

These lyrical phrases evoke symbols unique to Lovell, a small western town in north-central Wyoming: the domestic rose and the wild horse, both nurtured by human hands. The tone is at once a call to arms and a quiet reflection on the character of a local people. Those who forged a shield to protect these wild horses invited federal officials into the circle for conversation rather than litigation. If Lovell became a hotbed of activity, it was as an incubator for civil discourse instead of a standoff with the federal government, a paradox in this very "Red" state.

The Pryor Mountain Mustang is also unique: a wild horse with rare

Colonial Spanish origins. Although there has been some interbreeding with local ranch stock from the original homesteaders, genetic testing confirms this lineage. Wild horses are often associated with the western frontier, but here history dates back to the conquistadores who brought their small, sturdy mounts with them to Mexico in the late 1500s, then north to the US Southwest. The Crow Reservation in south-central Montana lies above the Pryor Mountain Complex, and some historians believe that the wild herd is descended from horses stolen or traded by the Crow that later escaped slaughter by the US Cavalry and slipped south into the forests of the Pryor Mountains.

Those who have spent time with Pryor Mountain Mustangs, both on the range and in adopted homes across the country, believe very strongly in making every effort to preserve their bloodlines. Respecting them as a wild (even native) species may be part of it, but there is so much more. Their Colonial Spanish physical type holds up a mirror to our precolonial past. These mustangs are smaller in stature than modern stock or pleasure breeds, but they seem to stand much taller: the illusion comes from their conformation as much as their spirit—necks and heads set high on deep, sloping shoulders and short-coupled backs capable of powering a collected "frame" that lifts the front end for an extended stride. A wild stallion posturing before a fight merits human, as well as equine, respect.

The ancestors of the Pryor Mountain Mustang were battle horses of a different sort, bred and trained for endurance, hardiness, and bravery. Many also have a unique trot or amble that is observable to the trained eye: a lateral trot—almost a running walk—associated with the Paso breed with which they share a genetic makeup. Although breeding these horses is a controversial practice among many activists who argue that they should be left wild and free, the fact is that, once adopted, Pryor Mountain Mustangs reveal an unusual willingness to form strong bonds with their human partners, as well as an uncanny ability to work with counselors in equine-assisted therapy. Testimonials are offered spontaneously and independently by many owners from around the country, and several are reported in this book.

The people of Lovell, Wyoming, were advocates for the Pryor Mountain Mustang long before passage of federal laws protecting free-roaming herds in western states. It is true that "Wild Horse Annie," Velma Johnston, fought to protect wild horses from the grisly practice of pursuing them in motorized vehicles and transporting them for slaughter. Nonetheless, the Pryor Mountain Wild Horse Range was the first public wild horse refuge, and the intent was to preserve an entire herd in large part because of its primitive Spanish markings and colors. The Reverend Floyd Schwieger, a Lovell pastor, kept a list and named every horse on the range. As this wily Lutheran minister was reported to have said, it would be harder to send an animal with a name to slaughter. He loved these mustangs, as did the small core of advocates who stepped up and closed the circle.

The BLM Targets Pryor Mustangs as Trespass Livestock

The exact origin of the Pryor Mountain Mustang is uncertain.[2] Today approximately 120–50 adult wild horses roam freely on the Pryor Mountain Wild Horse Range established by order of Secretary of Interior Stewart Udall on September 9, 1968.[3] For at least one hundred years prior to this Secretarial Order, horses roamed throughout the Pryor Mountains north of Lovell, Wyoming, and west of the Bighorn River. The Crow territory encompassed this area before the first homesteaders arrived in the mid-1800s, and Indians obtained thousands of horses through trade with other tribes. The horses roaming in Crow territory during the seventeenth and eighteenth centuries, like most Indian ponies, were likely descended from Colonial Spanish horses first introduced to the New World in the late 1500s by the conquistadores.[4]

The fate of the Crow Spanish horses was typical of Indian ponies in other tribes: many were captured and killed by federal government soldiers or homesteaders; however, a remnant likely survived. One theory about the origin of the Pryor Mountain Mustangs is that ranchers who were leasing reservation land north of the present-day Custer National Forest shot many Crow horses, pushing survivors southward. This core band of Spanish horses is most likely the foundation stock of the Pryor Mountain Mustangs.[5] These Spanish horses later mingled with homesteader horses turned loose on forest reserve and public-domain lands during the early 1900s. By the 1950s Lovell, Wyoming, residents familiar with these lands noticed a distinctive Spanish type among the free-roaming horses on East Pryor Mountain. Those residents formed the nucleus of advocates who fought to save the wild herd. Later blood typing and DNA analysis would confirm that many of these wild horses were of Colonial Spanish origin.

The Pryor Mountain Mustang is a wild strain of the Colonial Spanish type of horse first introduced to the United States via Mexico. The Colonial Spanish horse tends to be smaller and more compact than the European breeds imported by New England colonists. Present-day Spanish breeds are taller and bred for the elegance of movement required in dressage. By the late 1800s light saddle horses introduced by the Spanish conquistadores had all but disappeared from the United States, either shot or interbred with the larger European breeds. As noted by Colonial Spanish horse expert and doctor of veterinary medicine Phil Sponenberg, "These horses have Spanish conformation, and the blood types of the horses are also those expected of horses with Spanish ancestry. These horses are found along a major Crow and Shoshone migration route, and they probably have an origin in tribal horses. . . . The Pryor Mountain mustangs are an important resource for Spanish Horse conservation in North America."[6]

A core band of Spanish horses, the likely descendants of Crow Indian ponies, roamed the Pryor Mountains along with horses introduced by two groups of homesteaders: those who held grazing permits on twenty-four Pryor Mountain Forest Reserve allotments along the upper Sage Creek drainage, parallel to the southern boundary of the Crow Reservation, and ranchers south of the forest reserve boundary who homesteaded along Crooked Creek. Three hundred head from the reserve habitually migrated to lower elevations during the harsh winters, and the Forest Service took steps in the early 1900s to limit free-roaming horses, as their numbers began to exceed the carrying capacity of the area.

South of the Forest Service boundary ten settlers received homestead patents under the 1862 law. William Tillett was reported by a forest ranger to be running horses in the forest reserve. In an exchange foreshadowing confrontations between the Tillett family and the Bureau of Land Management during the 1950s, Tillett responded to a request to remove his horses by telling the ranger to remove the horses himself. The Forest Service built a boundary fence in 1937, but an unbranded wild herd continued to roam in the forest.[7] Rangers were unsuccessful in removing many of these wild horses. Charles O. Williamson recalled that among the free-roaming homesteader horses was a band of seventy head that were genuine little Spanish horses. He avoided gathering this band during the period when the Forest Service was trying to remove trespass horses.[8]

According to Hope Ryden, a feature producer for *ABC Evening News* who covered the Pryor wild horse controversy during the late 1960s, Williamson resigned his position with the US Forest Service because of the pressure put on him by the agency to get rid of the horses.[9] Spanish-type wild horse bands found roaming on the Tilletts' Bureau of Land Management allotment during the 1950s and 1960s were likely descended from the small band of horses protected by this forest ranger. Succeeding generations of free-roaming horses that were fenced out of the Forest Reserve were pushed eastward onto public domain that was part of the Pryor Mountain Complex. Until enactment of the Taylor Grazing Act in 1934, the public domain to the east and south of Custer National Forest was a livestock grazing "commons" open to all local ranchers. Once the US Grazing Service (now the BLM) drew allotment boundaries and issued grazing permits, the whole issue of trespass livestock came up again.

Lovell Advocates Urge the BLM to Create a Wild Horse Refuge

William and Bess Tillett owned TX Ranch, east of Crooked Creek and south of the national forest boundary. The name TX stands for the state of Texas, where Frank Strong, father of Bess Tillett, bought his first cattle and drove them north to Lovell, Wyoming, in 1895. A mixed herd of wild horses roamed on public-domain lands adjacent to TX Ranch, and these mustangs were used for ranch stock.[10] TX Ranch was later cited by the BLM for trespass violations because the horses roaming on its East Crooked Creek allotment were in excess of their permit for 20 horses.

The BLM advisory board for the Billings, Montana, Grazing District, elected by local permit holders, recommended in March 1950 closing a large area to horse use, extending west of Crooked Creek and south of

Custer National Forest to the Wyoming state line and east to the Bighorn River. For the next ten years, the trespass-horse problem on the south end of East Pryor Mountain was an ongoing subject of concern for the BLM. William Tillett's widow, Bess, appeared at a 1956 meeting of the advisory board to explain the use made by her horses in the East Crook Creek and Dryhead Common allotments. The BLM increased the TX Ranch permit from 20 to 40 horses, but in 1959 the BLM district manager sent her a letter reminding her that any number over that would be considered trespass horses.[11]

The trespass-horse situation that had spilled over onto BLM lands from the national forest continued to escalate during the 1960s, and TX Ranch was at the center of government efforts to reduce the estimated 150–300 free-roaming horses in the area closed to horse use in 1950. BLM officials considered any horses on cattle-grazing allotments trespass livestock, and permit holders were responsible for removing them.[12] Lovell-area ranchers with permits along Crooked Creek north of the Montana-Wyoming state line met with the Billings district advisory board on February 26, 1964, and agreed the number of trespass horses needed to be reduced; however, no one claimed ownership of the unbranded horses.[13] The term *trespass* applied to livestock left on the range in excess of the number allowed for an area.[14] According to Montana estray laws, unclaimed horses became property of the state. BLM officials, however, had legal authority to remove what it deemed to be trespass horses on its grazing allotments, and they put considerable pressure on the Tillett family to remove the horses. In November 1965 the BLM sent a certified letter to the Tillett family, warning of a trespass action.

On December 15, 1965, the Tilletts received a Notice of Trespass for "unlawfully allowing 132 more or less horses to graze in an area closed to grazing of horses." The advisory board was determined to resolve the situation and recommended that the BLM deny the Tilletts their grazing permit until they paid trespass fees and removed the horses.[15] The scope of the controversy expanded in 1966 when the BLM suspended the Tilletts' horse and cattle permits. The legal status of the wild herd was in limbo. Lloyd Tillett, son of Bess and William, claimed ownership of the wild horses roaming on their Crook Creek allotment in order to protect them from being removed.

When Lovell Chamber of Commerce members became aware of the impasse and intervened to help mediate the dispute, the parties reached a settlement. They believed that a wild horse herd would put Lovell on the map, attracting visitors to the area. On April 8, 1966, BLM district office officials met with the Tillett family and members of the Lovell Chamber. Thirty people attended that meeting, including the Reverend Floyd Schwieger, Howard Lusch, Chuck Wagner, and Phyllis Hill, who represented the historical committee of the Lovell Chamber. The Tilletts agreed to relinquish their claim to the wild horses on the East Crooked Creek allotment; the BLM, for its part, agreed to restore their grazing permits and to conduct an intensive study of range-forage conditions. While the study was under way, the horses would be left undisturbed.[16] The Lovell Chamber had emerged as advocates for the Pryor herd.

The Reverend Floyd Schwieger, a Lutheran minister in Lovell, acted as the mediator between BLM officials and Lovell advocates.[17] In a letter to Wyoming congressman Teno Roncalio, after the April 8 meeting he recounted local efforts to preserve the band of wild horses. He hoped to enlist the support of the congressman in maintaining a herd of horses roaming in an area on the south slope of East Pryor Mountain, adjacent to and including part of the proposed National Recreation Area (NRA). Members of the Chamber of Commerce historical committee saw good reason to preserve the Pryor herd. Wild horses had been in the area before 1900, when homesteaders first arrived. The horses were likely remnants of Crow horses from across the Pryor Mountains. The potential for tourism was excellent, since the Bighorn Canyon NRA would be on a main route to Yellowstone National Park.[18]

Members of the committee saw the Pryor wild horses as part of their western heritage and as a potential tourist attraction, but one issue plaguing their efforts was the question posed by both BLM and National Park Service officials: whether the Pryor horses were truly wild. The National Park Service superintendent for the Bighorn Canyon NRA had assured Lovell Chamber secretary Phyllis Hill that if the horses were truly wild, he would be interested in preserving at least a portion of the herd for visitor observation.[19] But after studying the issue, the NPS concluded that ranchers had manipulated the herd in the past, trying to breed up its quality to supply stock horses for working cattle.[20] The band of little Spanish horses identified by forest ranger Charles O. Williamson was represented on the range, but it would take several decades for scientists to identify them as a strain of Colonial Spanish horse.

Members of the historical committee were counting on tourists wanting to see a band of wild horses in a well-defined, accessible area. Even before the Colonial Spanish lineage of the Pryor Mountain Mustang became part of Lovell's plan for a wild horse visitor center, the close proximity of a wild herd was a novelty for the public. The earliest advocates had a vision that linked the town of Lovell to a nearby wild horse refuge. Their efforts began as early as 1967, while the BLM was conducting its study of the public range. They requested that the BLM and the NPS set up a wild horse refuge called "Yellowtail Horse Haven," with a controlled herd of eighty to one hundred horses on federal land south of Custer National Forest and east to Big Horn River, including part of the area proposed for a National Recreation Area.[21]

The Tilletts signed a "Waiver of Ownership of Horses" on October 26, 1967, relinquishing their grazing privileges on specific areas of their East Crooked Creek allotment, as well as on the Dryhead Common allotment south of Pryor Mountain and adjacent to the Bighorn River. The plan under discussion was to explore ways to clear those areas of cattle and to reserve the land for a herd of wild horses. The new district manager, Dean Bibles, concluded from the BLM range-condition survey, however, that it was necessary to cull a large number of horses from the wild herd.[22] In an April 5, 1968, letter to Montana senator Mike Mansfield, BLM associate director John Crow framed the dilemma facing the agency as choosing among alternatives for managing wild horses, big game, and basic soil and water resources. He noted a "wide spread of public opinion" and a series of public meetings planned to gather citizen input. The attached press release, dated March 14, referred to considerable public reaction and legal problems with the idea of a wild horse range. The BLM postponed action until after the public meetings.

The agency statement read: "Lovell, the town nearest the horse area, is interested in establishing a horse refuge in the Pryors as an additional visitor attraction near the Bighorn Canyon National Recreation Area." The BLM was considering the option of a refuge; however, agency officials warned of deteriorating water and soil conditions and proposed either retaining a small number or removing the entire wild herd.[23] The "public reaction" likely referred to big-game hunters, among others, since the alternatives presented in the attached press release mentioned balancing horse numbers against deer populations and reintroducing bighorn sheep to the area. The legal problems to which the statement referred had to do with ownership of the wild herd.

Dean Bibles was quoted as saying, "In order to legally establish a good horse refuge in the Pryor Mountain area, we would have to issue a grazing permit for a specific number of horses and a local group would then have to catch and brand the animals, plus conducting annual roundups to remove excess numbers."[24] Establishing a wild horse refuge within the existing federal and state legal framework required grazing allotments and permits for specific numbers of stock animals, state brand inspections and disposal of unclaimed or unbranded horses, and responsibility of the individuals or groups sponsoring the wild herd to remove excess animals if the population increased. The press release reflected the bind in which the BLM found itself, caught between public interest in the Pryor wild herd and the need to balance that idea with opposing interests as well as current legal constraints.

The BLM press release was a setback for Lovell advocates, and Chamber of Commerce members began to reach out to their congressional delegation, asking for pressure on the BLM to create the wild horse refuge with a viable number of horses. Elected officials from Wyoming and Montana were aware of the situation. Wyoming senator Gale McGee had earlier proposed to Secretary of the Interior Stewart Udall that the BLM complete a study of his proposal to annex a sixty-square-mile area of the Pryor Mountains to the Bighorn Canyon NRA.[25] A small band of wild horses already roamed in the Horseshoe Bend area at the entrance to the NRA and would be visible from a paved road proposed to connect Lovell and Hardin, Montana.[26] In May 1968 the Lovell Chamber explored the idea of having the wild horse refuge included in the Bighorn Canyon NRA. They planned to contact the Wyoming and Montana delegations, as well as the two state governors, the Wyoming and Montana Chambers of Commerce, and federal, state, and local agency officials.[27]

The mayor of Lovell, Cal Taggart, was an advocate for the Pryor wild herd, and he also worked with the Wyoming delegation. As the elected chief executive of Lovell, he believed that the long-term potential for the community as a whole outweighed the immediate concerns of those who were opposed to the idea of a wild horse range. While BLM grazing allotments on East Pryor Mountain were under the jurisdiction of the BLM, these public lands were adjacent to Lovell-area cattle ranches to the immediate south. Many area ranches had been owned by the same families for several generations, and ranchers considered their permitted allotments as an inheritance to pass down to future generations. The possibility of a wild horse refuge on East Pryor Mountain seemed to pose a threat to this tradition, although the family most immediately affected, the Tilletts, argued that the range was unsuitable for cattle and advocated for the wild herd roaming on their East Crooked Creek allotment.

Local ranchers were especially sensitive to possible BLM support of a public wild horse range, because another federal agency, the Bureau of Reclamation, had recently condemned and flooded farmland in Bighorn Canyon with the construction of Yellowtail Dam. Even though the federal government compensated landowners, the Bighorn Canyon National Recreation Area project was known as "the Take" by many Lovell residents. The flooding of cash-crop farms in the area now known as Bighorn Lake had a ripple effect throughout the agricultural sector of the local economy. Some implement dealers went out of business, as farmers moved their operations to nearby Powell, Wyoming. Thus, building political support for a wild horse range on federal land depended on the leadership of Mayor Cal Taggart. He faced resistance from local area ranchers who were distrustful of any federal government initiative that threatened their livelihoods.[28]

The mayor was also an advocate for a proposed transpark highway linking Lovell to Hardin, Montana. Influential families from Cody, Wyoming, worked quietly behind the scenes to halt this project, because it threatened to divert visitors who traveled through Cody to the east entrance of Yellowstone National Park. The proposed highway would have provided an alternate route through Red Lodge, Montana, to the north entrance of the park. Although Mayor Taggart and other Lovell advocates pushed hard for the transpark highway project, stronger political interests prevailed. The terminus of the new road ended up at Barry's Landing, below Yellowtail Dam. Visitors who wanted to see wild horses along that road would have to drive through Sheridan, Wyoming, and cross over the Bighorn Mountains into Lovell or nearby Greybull.[29]

The mayor used his considerable political influence to put pressure on the Wyoming congressional delegation for a wild horse range. The *Lovell Chronicle* was also a key source of support and influence, well beyond the immediate area. In March 1966 the *Lovell Chronicle* had published an open letter to Governors Clifford Hansen (WY) and Tim Babcock (MT) written by staff writer Beverly Robertson, with the title "Please Save the Horses."

> There has been a lot of horse talk in Lovell lately. Reverend Floyd Schwieger, a local minister, and Howard Lusch, Pacific Power & Light manager, have been urging people to write to you.
>
> Won't you use your influence in the preservation of a herd of about 140 head of wild horses that roam in rough terrain on the south side of the Pryor Mountains?
>
> Although they run mostly in Montana, the area is accessible mainly through Lovell and the new road built by the National Park Service off Highway 14A. It goes to the Big Horn Recreation Area.
>
> Rev. Schwieger says, "These horses are part of our western heritage, something that should be preserved for our children and grandchildren."[30]

As a result of this open letter, published and reprinted in newspapers across the country, hundreds of letters poured into Wyoming, addressed to Governor Hansen and to the Lovell Chamber of Commerce. The Chamber collected 187 individual letters written by schoolchildren, 537 petition signatures from children, 233 individual handwritten letters from adults, 137 signatures from adults on petitions, and 9 telegrams.[31] These personal appeals were reminiscent of the letter-writing campaign supporting the efforts of a Nevada woman to pass legislation prohibiting the practice of pursuing wild horses from motorized vehicles in order to sell them for slaughter. Her successful effort was bolstered by schoolchildren from across the country and made Velma B. Johnston (a.k.a. "Wild Horse Annie") a well-known advocate. In 1968 she joined the cause of saving the Pryor wild horses.

Velma B. Johnston was a well-known defender of wild horses across the western states. She first became involved in efforts to protect wild horses during the 1950s, when she worked with Senator Mike Mansfield, who had cosponsored the "Wild Horse Annie Law" in 1959, prohibiting the use of aircraft or motor vehicles to hunt wild horses on US-owned lands.[32] This law was the first legislation to provide any form of federal government protection for wild horses. She joined local efforts to save the Pryor wild herd in April 1968, corresponding with Schwieger and Mansfield. Her involvement brought national attention to Lovell because she was president of the International Society for the Protection of Mustangs and Burros (ISPMB). This intervention was the first, and only, time that an outside wild horse activist organization worked side by side with Lovell advocates.

In a lengthy letter to Senator Mansfield dated April 9, 1968, she called to his attention the long-standing controversy in Lovell between the BLM and Tillett family members who had a history of "endeavoring to protect this particular herd of wild horses" against the BLM's intent back in 1966 to reduce the herd to twenty head.[33] She pointed out that the BLM was once again proposing to practically annihilate the herd and asked for the senator's support, writing, "I am certain that if the lawmakers of the states involved will indicate their affirmative interest to the Secretary of the Interior, steps can be taken to bring about this and future settlements of an issue that will always be controversial just so long as the people are denied an equal voice with governmental bureaucracy."[34]

The Reverend Floyd Schwieger followed up this letter to Senator Mansfield with one of his own dated April 11, 1968, writing that he had visited the area where the wild horse herd roamed many times and observed, "This is as authentic a herd of wild feral horses as remains in our country. Their many colors and the oft appearing dark line back (dorsal stripe) attest to this fact." He acknowledged that there were "undesirable horses that have attached themselves to this band and should be removed," but he added that "Mr. Bob Brislawn from Oshoto, Wyoming, who owned several hundred pure Spanish Mustangs, says in a letter that according to their markings there may be considerable Spanish blood in these horses." Schwieger emphasized, "We are not asking that wild horses be established on the other plentiful wildlife areas." And he noted that "incidentally domestic animals do not and cannot inhabit the area because of a lack of water." He also asked Senator Mansfield for his help in protecting the Pryor wild herd.[35]

Wild Horse Annie wrote to Floyd Schwieger on April 24, 1968, that she had received responses from Senators Mansfield and Metcalf of Montana. "Judging from Senator Mansfield's letter," she commented, "quite a bit of ruckus has already been stirred up, and when the more than 500 directives of the ISPMB are received by members and friends, I am sure there will be a deluge of letters to Washington addresses."[36] She remained involved with Lovell wild horse advocates even after creation of the Pryor Mountain Wild Horse Range, by working with Schwieger to sponsor the adoptions of young horses airlifted from the top of East Pryor Mountain by the BLM in 1970 in order to protect them from starvation due to harsh winter conditions. This partnership came to an amicable end, and Lovell advocates chose to continue their work without help from outside wild horse activist organizations.

Lovell Advocates Propose to Sponsor the Wild Herd

Members of the Lovell Chamber historical committee formed the Pryor Mountain Wild Horse Association (PMWHA) in June 1968 in order to sponsor the herd and maintain at least seventy-five to one hundred wild horses. Charles Wagner was the president, and Floyd Schwieger was vice president, while the Chamber's secretary, Phyllis Hill, also served as secretary of the PMWHA. The group hoped to find a solution to the escalating controversy by working with Dean Bibles, the new BLM district manager. In a letter to Wagner, Bibles spelled out the requirements for a sponsor: purchase and brand the horses, and conduct annual roundups for branding foals and culling excess animals.[37] Although Wagner was working to build support from national organizations such as Wild Horse Annie's ISPMB and Tom Holland's National Mustang Association, he stressed to the media that he was trying to cooperate with the BLM to see what could be worked out.[38]

A feature producer for the *ABC Evening News*, Hope Ryden, made several trips to Lovell during June and July 1968 based on a tip from John Walsh, special agent for the International Society for the Protection of Animals.[39] She established a working relationship with Charles Wagner and made trips to the most inaccessible parts of the range with Lloyd Tillett, taking a large number of photographs on both BLM and National Forest Service lands.[40] Her reporting revealed that national BLM director Boyd Rasmussen was unaware of the PMWHA offer to sponsor the herd, and she notified Wagner. He planned to follow up with a letter to the BLM director, pledging to sponsor the wild herd and provide supplemental feeding during the winter months. The BLM district office, however, continued with plans to gather the herd.[41]

The *ABC Evening News* aired Hope Ryden's story in which she reported that BLM staff were about to trap and gather the wild herd. In a handwritten note, Lloyd Tillett wrote to her on August 5, "Well the boys of the BLM are busy building the horse trap and fence, it looks like they are about half way through I will let you know when the trap and fence are complete and when they start to trap the horses."[42] In the meantime, however, Wagner wrote to her about a meeting with Dean Bibles in Bridger, Montana, in which Bibles informed Wagner that he was leaving for a highly confidential meeting in Washington, DC, on August 5. Dean Bibles was reaching out to Lovell advocates to get a sense of their support for a proposal that would set up a refuge under BLM jurisdiction. The BLM was dropping its requirement for a local sponsor to own and manage the herd and was proposing at least sixty horses. The Montana Livestock Commission had agreed to this proposal, allowing a BLM brand on the upper lip of each horse to substitute for the state brand.[43]

The BLM proposal called for range improvements and the potential for greater numbers of horses. In his letter Wagner reported that he and Floyd Schwieger were careful about how they worded their response to Bibles. "We did not wish to be quoted by BLM as agreeing to anything because we actually weren't agreeing to details only to his effort to reach a satisfactory (for the horses) solution." Floyd Schwieger had called Velma Johnston, and she thought the Lovell group should hold out for one hundred horses. "We want to," Wagner wrote, "but it will require findings of an independent range study to show that 100 horses could be supported." In a handwritten note on the side of his letter, he added, "P.S. I really think Bibles and superiors are searching for an answer. Hence we have agreed to keep this latest development out of the press for the time being."[44]

In the middle of August 1968, Wyoming representative William Henry Harrison announced that the BLM had notified him of a reprieve for the wild herd due to the Interior Department's plans to form a long-range wild horse policy and to give the Humane Society of the United States (HSUS) time to complete an independent range study. The announcement was based on a conversation between legislative staffer Douglas Baldwin and BLM employee Niles Humphrey. On August 12, however, BLM assistant director Jim Crow advised the congressman by letter that the BLM had yet to make up its mind about the option of rounding up most, or all, of the wild herd.[45] He called Harrison's announcement a misunderstanding. The congressman wrote a letter to BLM director Boyd Rasmussen, urging him to accept the PMWHA offer to sponsor a wild horse refuge; however, his staffer wrote to Hope Ryden that he was furious about the BLM letter. According to her notes, he stated that "they had assured us without equivocation that there would be no trapping of unbranded horses for a year or so."

Velma Johnston sent the Reverend Floyd Schwieger excerpts from a letter written to her by Hope Ryden: "As you know the BLM reneged on their agreement to give the unbranded horses a year's grace. So if they are in the mood to let that kind of embarrassing situation develop (deny what they agreed on to Congressman Harrison's office after the Congressman had sent out an AP [Associated Press] and UP [United Press] release) I can't imagine that our horses stand a chance." Wild Horse Annie's letter went on to say that she had gotten a phone call from the HSUS, which was expected to file a lawsuit against the BLM.

She took Schwieger's advice to avoid calling in humane organizations for help, but the widespread publicity had likely stirred up HSUS members. She referred attorneys and officials to Charles Wagner, as president of the PMWHA, and ended her letter by asking Schwieger to "please include me in your prayers for the wild ones. I am so tired, so very tired."[46]

Secretary Udall Creates the Pryor Mountain Wild Horse Range

Political pressure on the BLM mounted during the next several weeks. A legislative assistant to Wyoming senator Clifford Hansen wrote to BLM director Rasmussen, asking for a memorandum defining federal and state government jurisdiction over the horses, a copy of the BLM range survey leading to the March 14 press release, and reassurance from the BLM that it would take no action until the HSUS completed its study.[47] The Tilletts communicated to Ryden that the BLM had completed a corral trap. On August 22, 1968, the HSUS sent a registered letter to Secretary of the Interior Udall, asking for reassurance that the BLM would take no action against the Pryor Mountain wild horses until it had made public its management plan. When the HSUS had received no reply after several days, the organization filed suit against the BLM.[48]

The HSUS and coplaintiff Lloyd Tillett sought a temporary restraining order to prevent the destruction of the mustang herd. BLM attorneys argued in DC federal district court that the agency had no plan to trap and dispose of the wild herd and that once it had formally announced its decision, there would be ample time for all to be heard. The federal judge denied the plaintiffs' request for a temporary restraining order, but left the HSUS complaint pending before the court.[49] Two weeks later BLM director Rasmussen announced that the interior secretary had set aside thirty-one thousand acres in the Pryor Mountains as a wild horse and wildlife range. The range available to wild horses included lands within the Bighorn Canyon NRA. Wyoming senator Gale McGee, who had supported efforts of the Lovell Chamber of Commerce, praised the decision.[50]

Nearly twenty years later, Hope Ryden argued that the HSUS was influential in the final decision:

In 1967 and 1968, I became aware of a Bureau of Land Management plan to drastically reduce or totally eliminate the Pryor Mountain herd. The three options offered were as follows: remove all horses, remove all but 15 horses or remove all but 35 horses. None were acceptable by local people, who protested that all three options would result in extermination of the herd. My coverage of this plan was aired on ABC-TV News and created quite a stir. In the end the Humane Society of the United States took the matter to court and an embarrassed Secretary of Interior, Stuart Udall, designated the Pryor Mountains a Wild Horse Range.[51]

Secretary Udall's designation of the Pryor Mountain Wild Horse Range was dated September 9, 1968, and filed with the *Federal Register* on September 11, 1968.[52] This wild horse range was the first refuge established by the federal government and open to the public. Senator Mike Mansfield from Montana entered the following BLM news release about the designation into the *Congressional Record:* "Secretary of Interior Stewart L. Udall has set aside a 31,000 acre wild horse and wildlife range in the Pryor Mountains along the Montana-Wyoming border, Bureau of Land Management Director, Boyd L. Rasmussen announced. The move is being made to give Federal protection to a herd of wild horses whose future has aroused nationwide attention for months. . . . We all recognize that these horses are a national asset, and the Pryor Mountain herd is suitable for management of the wild horse so important in Western history."[53]

The Secretarial Order stated, "Subject to valid existing rights the area will be primarily administered for the protection and management of wild horses, wildlife, watershed, recreation, archeological and scenic values." The range encompassed BLM lands on the southern slope of East Pryor Mountain, including the former East Crooked Creek cattle-grazing allotment north of the Wyoming-Montana state line, as well as land to the east ending at the Bighorn Canyon. Secretary Udall assigned responsibility for developing a wild horse management plan for public lands within the range to the BLM and, in cooperation with the National Park Service, for lands within the new Bighorn Canyon National Recreation Area.[54]

After signing the declaration, Secretary Udall authorized BLM director Rasmussen to appoint a special advisory committee to recommend suitable methods of balancing wild horses, deer, and food available for them.[55] The advisory committee included Velma B. Johnston and Clyde Reynolds, Lovell's current mayor, but Lovell advocates were keenly aware that creation of the PMWHR was just a prelude to long-term preservation of the Pryor Mountain wild horses. Lovell advocates had to convince committee members to make recommendations to the BLM consistent with their vision. The Reverend Floyd Schwieger spoke at a hearing before the advisory committee in Lovell about the Spanish lineage of the Pryor horses, and PMWHA secretary Phyllis Hill underscored the value of a wild horse herd within the Bighorn Canyon National Recreation Area.[56]

Earlier in the year the BLM had begun staking a fence to enclose the estimated two hundred wild horses on public rangelands.[57] A portion of the fence ran along the boundary of the Tilletts' property. A gate between the PMWHR and the TX Ranch was left open, allowing unrestricted movement of the horses. By one account the BLM gave Lloyd Tillett six months to claim "his" horses and move them to TX pastures, and Tillett wanted the blacks, browns, and tiger-striped buckskins.[58] Once the fence was closed, fifty to sixty horses were on Tillett property, and an agreement with the BLM gave him title to these Pryor horses.[59] The Tilletts' Pryor Mountain Mustangs would later become one of two sources of foundation stock for breeders who wanted to establish "captive" herds in order to conserve the gene pool of Pryor Mountain Mustangs. These breeders considered any Pryor horse certified by the Tillett ranch to be eligible for registration.

The special advisory committee held a press conference at the federal building in Billings in late March 1969 to report on its initial recommendations to the BLM. Highlights of those recommendations included making every effort to retain at least one hundred horses on the range, as well as selection of animals for herd quality based on coloring and physical characteristics found in the Pryor Mountains.[60] The Spanish "type" was essential to the vision of the PMWHA. Their reasons were twofold: Floyd Schwieger had spent numerous hours observing the wild herd, and he was the first to notice distinctive primitive marks, such as zebra stripes and line-back (dorsal) stripes on the dun- and mouse-gray-colored horses. He had a passionate interest in

Spanish Mustangs and in the preservation of the Pryor Mountain wild herd. The second reason was the potential tourist attraction.

The Lovell Chamber of Commerce hosted members of the special advisory committee and a number of BLM officials at a dinner during their four-day meeting during May 1970. Lovell mayor Clyde Reynolds reported that committee members were satisfied with the condition of the horses when they took a trip to the range the next morning.[61] Efforts by Lovell Chamber members and the PMWHA to save the wild herd appeared to have succeeded, due in part to their persistence in working with BLM officials. PMWHA president Charles Wagner had reflected this philosophy in June 1968 when he stated that he was trying to act with all cooperation with the BLM to see what could be worked out to save the wild horses. And Floyd Schwieger had advised his friend and ally Wild Horse Annie to avoid calling in humane organizations for help.

Local opinion about the "Lovell way" of working with the BLM was mixed. While some local people may have protested the wild horse removal plan, other residents resisted the idea of transforming Lovell from a cattle town into a tourist destination. Opponents dismissed the herd as a bunch of old ranch horses turned loose on the range. Lovell advocates had to make a case for the unique Spanish lineage of the herd based on their observations of conformation, color, and markings, since the availability of genetic analysis was decades in the future. To make matters more complicated, the late 1960s was a time when concerns about the environment were on the rise. Federal natural resource agencies were under considerable pressure from Presidents Kennedy and Johnston to protect the national forests, parks, and public rangelands. Wildlife advocates viewed wild horses as a feral or nonnative species that returned to the North American continent in the late 1500s with Spanish explorers after they had long since become an extinct species on this continent. From their perspective, only small grazers such as rabbits and pronghorn that evolved within this fragile semiarid ecosystem should remain in the area.

In October 1971 the *Lovell Chronicle* reported: "Here, where the first wild horse preserve was proudly presented to the public two years ago, a bitter battle is again shaping up, with some new arguments thrown in just to keep things interesting." The special advisory committee was wrapping up its three-year review and held a closed session with environmentalists opposed to having a substantial herd of wild horses on the newly designated PMWHR. The PMWHA and the Lovell Chamber were asking for a thorough study of range conditions before removing any wild horses, but the advisory committee excluded them from the closed session.[62]

Lovell advocates were allowed to speak at a public hearing, pointing out that the proposed NRA highway through a portion of the PMWHR would have a minimal impact on movement or forage for wild horses and questioning what criteria the BLM would use to reduce the number of wild horses. Charles Wagner, PMWHA president, argued that the BLM and special advisory committee needed to complete a thorough study to provide criteria for removals. The position of the PMWHA was that if a removal were to be deemed necessary, strains of Spanish horses should be retained. He pointed to the skeletal remains from Forest Service and BLM rangelands that demonstrated the unique nature of the Pryor wild horses. The number of lumbar vertebra was smaller in wild horses of Colonial Spanish descent versus descendants of stray homesteader horses.[63]

By the middle of November 1971 the BLM decided that, based on the data collected, none of the foals would survive the coming winter. Wranglers roped the foals and prepared each one to be airlifted from the range to Lovell. A photo taken by the *Lovell Chronicle* shows BLM wrangler Lynne Taylor on horseback carefully holding the lasso around a foal to prevent the noose from tightening around its neck. The foal is staring at the camera, but stands calmly next to Taylor. A total of sixteen foals were airlifted to Lovell, while an additional twenty-nine stallions were given to Crow families interested in owning a Pryor Mountain wild horse. A previous BLM census had estimated the herd ratio as 60 percent stallions to 40 percent mares and wanted to return to a more natural balance. The Reverend Floyd Schwieger supported that strategy, pointing out that it would help to preserve the Spanish mustang bloodlines.[64]

The foals spent the winter in Lovell under the care of local veterinarian Ed Lowe and the sponsorship of Wild Horse Organized Assistance

(WHOA), a new organization headed by Velma B. Johnston, with Floyd Schwieger serving as the local agent. The consensus in Lovell was that if culling was necessary, then the present plan was the most acceptable.[65] In a statement to the *Lovell Chronicle*, Wild Horse Annie said: "I hope it is clear to everyone why we're doing this." Foals would be saved from almost certain death during the upcoming winter, while the mares would have a better chance of survival without the stress of lactation. Her organization would screen adoption applications and ensure that each foal ended up in a good home.[66] By March 1972 the foals were ready to go to their new homes, and Wild Horse Annie published a heartfelt thank-you in the *Lovell Chronicle*.

Her thank-you extended to Floyd Schwieger for coordinating various phases of the operation, to Dr. Lowe for ensuring healthy foals and for gentling them before they met their new owners, to the BLM field and administrative personnel who went above and beyond their assignments to carry out the removal in a humane fashion, to the Lovell Chamber and Lovell residents for their warmth and hospitality, and to the *Lovell Chronicle* for generous coverage of the months-long effort. Her eloquent concluding thoughts are worth quoting in full:

> In future years when those who chronicle our country's history and those who write its legends record the glowing chapter of the fight for the Wild Ones, it will be noted that four FIRSTs occurred in or near Lovell:
>
> The Pryor Mountain Wild Horse Range is the FIRST such range designated in America to which the public has access.
>
> In October, 1968 the FIRST committee ever appointed to represent all of the interests involved in the use of our Public Lands met in Lovell.
>
> In November, 1971 a reduction in numbers of wild horses on the Pryor Range was made. It was the FIRST time that any reduction, either official or on an individual basis had been conducted with the primary concern being the humane treatment of the horses.
>
> The Pryor Mountain colts will be the FIRST ever to go forth with a specific identity in lieu of the limbo that has been the fate of their kind since the first mustang sought and found freedom in the lush grasslands of early America.[67]

A Legal Framework for the PMWHR Changes the Landscape

The PMWHA balanced its approach between asking members of the Montana and Wyoming congressional delegations to put political pressure on BLM officials in Washington, DC, while at the same time signaling to local BLM field office officials that they understood the agency's mandate to be stewards of public rangelands and were open to consultation about the number of wild horses that should remain on the range. Relationships with the Billings office were based on trust. Efforts to avoid adversarial relationships were the rule, rather than the exception, as the BLM built the legal scaffolding to preserve the wild herd while protecting fragile rangeland ecosystems. This consultative approach was difficult nonetheless, because Lovell advocates and BLM field office personnel had very different views about the wild herd.

Disagreements about the origin of the Pryor wild horses had surfaced many years earlier. The National Forest Service and the BLM viewed the free-roaming horses as strays, escaped or released by ranchers in the area. Lovell advocates, on the other hand, noticed distinctive colors and markings suggestive of Spanish descent. A key to preserving the unique strain of Colonial Spanish horse that came to be known as the Pryor Mountain Mustang was retaining stallions and mares on the PMWHR that were of a distinctive "type" and then keeping records of family lines. BLM wrangler Lynne Taylor kept many early records, and then the Reverend Floyd Schwieger took extensive notes and numerous photos over a period of years. Despite differences in viewpoint, mutual trust paid off when the BLM incorporated preservation of the Pryor wild horse type in its first herd management plan.

Most significant for public management of the PMWHR was the *Herd Management Area Plan* for the PMWHR issued in 1984, amended in 1991, and revised in 2009. *Herd Management Area Plans* were required for all wild horse herds following passage of the 1971 Wild Free Roaming Horses and Burros Act as the administrative scaffolding supporting actions to improve range conditions and support healthy herds. Plans for the PMWHR were unique in selecting for characteristics of the Pryor Mountain wild horse, in addition to other criteria such as age. In later years periodic removals and fertility control were part of ("tiered to") the 2009 *HMAP*. Lovell advocates were able to assist in monitoring of the range and wild herd, identifying family bands, recognizing distinctive Spanish characteristics, and making recommendations to the BLM about which horses to keep and which to remove. This consultative effort became a model of twenty-first-century public management, developed over many years.

In 1978 the Reverend Floyd Schwieger moved his Lutheran ministry to another Wyoming community. A letter from the BLM Billings Field Office underscored his previous contributions:

> Your assistance and understanding of management issues and problems have helped promote a deeper public understanding of wild horse biology and habitat requirements. Back in the mid-1960s, the fate of the Pryor Mountain mustangs in the canyon lands west of the Bighorn River was in doubt. The controversy which surrounded this management issue was finally resolved with the designation of the nation's only wild horse range.
>
> Your efforts in helping BLM establish the Pryor Mountain Range were significant. Throughout that period of turmoil and emotionalism, you never lost sight of the basic land resource issues involved. Your efforts gave a boost to the Bureau's ability to initiate responsible management programs in the Pryor Mountains.[68]

Many of the earliest advocates had retired or passed away, while a new generation would emerge to take up the reins of the effort to preserve the Pryor Mountain Mustangs. During this period the BLM developed the framework for managing the PMWHR, including expansion of boundaries and additional acreage through lease arrangements with private landowners. Nevertheless, the PMWHR was an enclosed area with a large portion at the base of East Pryor Mountain being semiarid, and sufficient quality forage and perennial water sources were in short supply. The PMWHA would form again in 1989 and establish a long-term vision to preserve the Pryor Mountain Mustang.

Chapter Two

Growing Pains

Expanding the Pryor Mountain Wild Horse Range and Lovell Advocacy Groups

The first several years following designation of the Pryor Mountain Wild Horse Range were critical for establishing physical and legal boundaries and making key decisions about the size and characteristics of the wild horse population. An experienced BLM wild horse wrangler estimated that the population before the 1971 gather was more than two hundred, including the fifty to sixty head on the Tilletts' property. Billings district manager Rex Cleary and Dr. Wayne Cook, chair of the special advisory committee, offered a cautiously optimistic assessment that native plants on the range would come back if given a few years of respite.

The New Wild Horse Law Authorizes Expansion of the PMWHR

The original boundary of the PMWHR encompassed 33,699 acres of BLM and NPS lands in the state of Montana. In 1969 a second Secretarial Order made another adjustment, adding lands within Wyoming.[1] The BLM lands included the East Crooked Creek allotment on the southern slope of East Pryor Mountain that the Tillett family had officially relinquished in 1967, along with a waiver of ownership of the wild herd. The Tilletts also released their grazing rights in the Dryhead Common cattle allotment that became part of Bighorn Canyon National Recreation Area in 1966. The wild herd was fenced out of National Forest Service lands, as well as from the Mystic allotment that surrounded private land and a parcel of state land, both at the top of East Pryor Mountain.

The 1968 Secretarial Order called for the BLM and NPS to develop a management plan for the range, but the BLM operated without statutory authorization or appropriations for the first several years. Because the PMWHR predated the Wild Free Roaming Horses and Burros Act by three years, the legal framework governing public management of the wild horse herd was a prototype in many respects.

The 1968 Secretarial Order stated: "The Pryor Mountain Wild Horse Range shall be in all respects subordinate to the Bighorn Canyon National Recreation Area established by the Act of October 15, 1966 so far as it affects lands comprising any part of the Bighorn Canyon National Recreation Area. Wild horses within the area shall be managed by the Bureau of Land Management, in a manner that is compatible with the purposes for which the Bighorn Canyon National Recreation Area was established."[2]

This provision gave the National Park Service considerable control over PMWHR lands within the NRA. The Dryhead, as the area came to be known, was the most accessible to the public because of the low elevation and the construction of a paved road running north to Barry's Landing along the Bighorn River. The NRA was bounded to the west by Sykes Ridge, although wild horses could easily cross over the rocky ledges to the southern slope of East Pryor Mountain. The Lovell Chamber of Commerce had supported the existence of a wild herd in the NRA as a tourist attraction, and a substantial number of horses already roamed in the Dryhead.

During the next twenty-year period, the NPS agreed to a temporary extension of the range north of the original boundary within the NRA, but the "Sorenson Extension" is no longer part of the range. The NPS monitored conditions over the years and decided that concerns about overgrazing and negative impacts on wildlife justified canceling the agreement in 1990. The NPS has been supportive of having a small herd of wild horses, and park rangers have worked cooperatively with the BLM on removals, as well as on fertility control, "darting" mares that are identified by BLM field office personnel. Congress established the NRA, however, to preserve the area's cultural and natural resources, including wildlife. The NPS views the wild herd as part of the area's cultural heritage and recognizes its value as a tourist attraction, but views the wild horse as a nonnative species in potential conflict with wildlife. As a result the BLM reduced the population level of the herd for a period of time beginning in 1992 to compensate for boundary adjustments made by the NPS.

The boundary of the PMWHR also expanded to include National Forest Service lands after passage of the WFRHBA. The 1971 law expanded federal protection to wild horses on public lands throughout the western states, delegating responsibility to the secretaries of the interior and agriculture. In 1972 BLM wildlife biologist Ron Hall conducted a comprehensive assessment as part of a joint land-use planning process. One of the major objectives of this study was to examine the expansion of existing boundaries of the PMWHR in order to return the wild horses to areas of historical use. The 1971 law provided that "wild free-roaming horses and burros shall be protected from capture, branding, harassment or death; and to accomplish this they are to be considered *in the area where presently found,* as an integral part of the natural system of the public lands."[3]

Small bands were known to be roaming outside of designated boundaries enumerated in the Secretarial Order, and the intent of the new law was to establish a herd area where wild horses roamed. It delegated authority allowing the BLM to explore the potential of expanding range boundaries into lands administered by the US Forest Service in the Department of Agriculture, as well as leasing private and state lands. The Hall study concluded that several areas within the Pryor Mountain Complex probably had wild horses on December 15, 1971, the effective date of the WFRHBA. The first area was the Mystic grazing allotment, near the top of East Pryor Mountain, where horses were observed to have wintered during 1971–72. A fence along the northernmost boundary of the range was in poor repair, and a small band of horses regularly crossed into this allotment. The study noted that the area had adequate amounts of water and forage and that it would provide both winter and summer range for horses.

The second area lying outside the range was the Lost Water Canyon area within Custer National Forest (also known as Area Q), encompassing 4,000 acres of grassy ridges separated by timbered canyons. A small band had wintered in this area as well, and Hall made a point of saying that foals in this band had the highest survival rate during the winter of any area on the current range. The third area used by wild horses was a 7,000-acre extension north of the range boundary along the Bighorn Canyon that included the Sorenson Ranch. The NPS had jurisdiction over this part of the National Recreation Area. The fourth, and final, area identified by the Hall study was the remaining portion of the Crooked Creek allotment contiguous with the Tilletts' property line. The study noted that the horses owned by Lloyd Tillett and separated from the wild herd by a boundary fence were of the same genetic background.[4]

The study recommended that the PMWHR be expanded to include these, as well as other, areas known to be used by wild horses as of December 15, 1971. The BLM and US Forest Service reached an agreement in 1974–75 to expand the PMWHR to include the four areas identified in the Hall study. The Mystic and Crooked Creek areas were submitted to the secretary of the interior for formal addition to the range, while the Lost Water (USFS) and Sorenson (NPS) areas were proposed as informal additions, to be evaluated within five years.[5] The first *Herd Management Area Plan* for the PMWHR approved by the BLM, USFS, and NPS in 1984 documented 36,600 designated acres and 10,211 undesignated acres, for a total of 46,811 acres. The additional designated acres came from the Crooked Creek allotment, while the new undesignated areas included Lost Water Canyon within Custer National Forest, the Sorenson Extension area of the Bighorn Canyon NRA, state of Montana leased lands, and Mystic allotment leased lands.[6]

The Hall study pointed out that, in the past, wild horses had roamed freely within the entire Pryor Mountain Complex. Long before 1971, however, the USFS had erected a boundary fence, preventing horses along Crooked Creek from moving north onto forestlands during the spring and summer seasons. While a band likely remained inside USFS boundaries, most horses ended up on public-domain lands that became the PMWHR. As Hall concluded, man eliminated wild horses from most of the areas within the Pryor Mountain Complex years ago, and although parts of the Custer National Forest were historical range of the wild horse, they were excluded by law from consideration as part of the expanded PMWHR. The expanded boundaries became part of the Pryor Mountain Wild Horse Range.

The Hall study described the Pryor wild horses as a mustang type that was small, thick-necked, short-backed, and tough enough to survive the harsh conditions on the range. He wrote, "The mustang was a horse that was selectively bred by nature over time. He was a unique product carved and molded to fit the land." Even though homesteaders turned their domestic stock loose, adding new blood to the wild herd, the natural selection process led to certain mustang-type characteristics throughout the herd. "In a situation like the Pryors where the diversity of types of horses has been present, the selective process of nature continually provides the type of horses most capable of surviving. As a result, these horses have been gradually reverting back to the original mustang type horse. Evidence of this is the small size, the tough thick hoof wall, the reversion of colors to the blue corn, buckskins and bays with the line stripe down their back, along their withers and on their legs, and many other factors."[7]

The study also pointed to the many skeletons found on the range with a condition of fused lumbar vertebrae characteristic of a Colonial Spanish–type horse, although there were other lumbar vertebrae typically found in quarter horses. The colors of horses observed in the study were bay, black, brown, sorrel, red roan, blue roan, grullo, blue corn, and buckskin, but not palomino, pinto, or appaloosa. Twenty years later scientists working with the BLM and the PMWHA confirmed the presence of a Colonial Spanish horse type within the herd, reinforcing the findings from the Hall study.

The Hall study also addressed the issue of population control versus no control by referencing recent studies of density-dependent self-regulation by wild animal populations. There could be no "balance of nature" in the PMWHR, according to Hall, because wild horses were living in a limited area with declining range conditions. Disease and starvation were the major factors affecting population increases within the wild herd. Even if the BLM were to accept the effects of low nutritional status, such as a reduced breeding period and life span, as additional regulatory factors, it would have to wait years for the population to stabilize. In fact, the BLM took the position that removals were the most effective and humane approach in order to balance rangeland and wild horse health. The Hall study concluded, "The soil resource has been seriously depleted in past years. If this situation continues, the number of horses that can be supported in the Pryors will continue to decline. In essence, if the philosophy of allowing horses to seek their own population level is adopted, we will ultimately destroy the recreational resource we are attempting to perpetuate, the horse."[8]

The BLM and USFS Joint Agreement signed in 1974 decided that annual removals were necessary to maintain a balance between wild horse numbers and available land area, even with the range expansion from the original 33,600 acres to 36,600 designated acres and an additional 10,211 undesignated acres from agreements with the NPS, USFS, state of Montana, and private property owners. The decision was to remove only young animals when possible, while retaining enough young horses on the range to stabilize the age structure.

The Joint Agreement recognized the need for conformation selection criteria in order to preserve the unique characteristics of the Pryor Mountain Mustang. The decision was for the BLM to develop a conformation guide as part of its first management plan. Background analysis for this decision discussed mustang conformation guides published by various organizations, including the National Mustang Association, but identified the special prominence of the Pryor Mountain Mustang as a reason for designing a separate conformation guide. As it turned out, these horses are a unique strain of Colonial Spanish mustang, a critical conservation priority of the American Livestock Breeds Conservancy.[9]

The first *Herd Management Area Plan* in 1984 estimated that the horse population was 120–30 head following a 1971 removal. Foal crops brought the number back up and led to another removal in 1975 to bring the population back to 120–30 adult horses at the same time additional land areas became available to the herd. There was another removal in 1977, followed by a severe population reduction to 70 head after the winter of 1977–78. The herd numbers rebounded to 180 head, including the foal crop, resulting in two additional removals before publication of the 1984 *HMAP*. The herd population (one year and older) at the time of the first management plan was 129 head, with 70 percent under nine years old and only those horses considered the best in color and conformation within each age class left on the PMWHR following previous removals.[10] The range management objective of the 1984

HMAP was to "maintain a viable breeding herd which could perpetuate the characteristics of the Pryor Mountain wild horses."[11] In the years to come, the BLM and Pryor Mountain Wild Horse Association gained new knowledge about the genetics and phenotype (that is, physical characteristics) unique to the Pryor Mountain Mustang.

During the years immediately following enactment of the 1971 act, Dr. Jay Kirkpatrick, a reproductive physiologist, conducted a fertility study for the BLM. He would later become the director of science and conservation at Zoo Montana and develop the fertility vaccine known as PZP that proved to be a significant alternative to removals in managing population growth. As he recounted his research in the PMWHR, Dr. Kirkpatrick noted that "the agency foresaw the rapid increase in wild horse populations that would take place over the next decade and the need for publicly acceptable methods of controlling these animals."[12] During his fertility study he noticed unique characteristics in the Pryor wild horses. They tended to be smaller than other wild horses, had very thick hoof walls, and had structural variations such as five lumbar vertebrae in comparison to six in other domestic and wild horses. Their colors were typical of primitive-type horses: no whites, palominos, or pintos, and there was a lot of striping on their legs and backs.[13]

Most traces of Spanish mustangs were gone, according to Dr. Kirkpatrick, and the wild horses in the West were descended from horses turned loose during the Great Depression. He suspected that a few Spanish horses survived and evolved in the isolated Pryor Mountains. He noted that there was documentation showing horses in the area for at least two hundred years, but not enough time for the unique characteristics he had discovered to be due to evolution.[14] In short, the unique structure or phenotype found among Pryor wild horses was likely related to Colonial Spanish bloodlines. His hypothesis was supported in the early 1990s through genetic analyses conducted by Dr. E. Gus Cothran from the University of Kentucky.[15] Preservation of the Pryor Mountain Mustang as a unique Spanish type became the primary vision of the PMWHA.

The BLM introduced a small number of wild horses from outside the PMWHR during the late 1980s. This strategy became very controversial, and the agency eventually removed those horses and their offspring from the range during the early 1990s. The manager of the Billings Resource Area, Bill McIlvain, explained the rationale for transplanting a band from the Kiger Gorge range near Burns, Oregon: "Our evaluation of the Pryor herd in recent years indicated that inbreeding could become a problem before too long. So we started looking for a few select horses that could be added to the herd—horses with characteristics similar to those traditionally found in the Pryors."[16] The BLM also moved several wild horses from Wyoming herd management areas to the range.

The size of the herd remained at approximately 120 adult horses during this period in order to balance horse numbers with available forage. The distribution of the herd, however, was an issue, because BLM monitoring showed that the range topography naturally divided the horses into three subherds of 25–50 animals. Those subherds included the Dryhead area in the Bighorn Canyon NRA, the Tillett Ridge area north of the Tillett property, and Sykes Ridge, which formed a barrier to the west of the Dryhead. The 1984 *HMAP* documented a tendency for the Dryhead horses to remain in that area throughout the year, while horses from the Tillett and Sykes Ridge areas migrated during the spring and summer up to the top of the mountain, where the two subherds intermingled. The BLM's concern was the small number of breeding-age adults within each subherd. Later genetic studies would show a sustainable level of genetic diversity within the Pryor herd, but during the 1980s the BLM had to base its decisions on observations of range use and physical characteristics.

The 1984 *HMAP* stated as its planned action that the BLM would "continue to select horses for excess keeping those horses with the better conformation and enough color so there is somewhat of a balance."[17] The BLM justified adding Kiger horses to the Pryor Range because of the Spanish heritage of that isolated herd; however, this approach became the subject of much discussion with the Pryor Mountain Wild Horse Association during the 1990s. The new Pryor Mountain Mustang Breeders Association was especially critical of the outside introductions. Both organizations worked with the BLM to remove as many of those horses as possible from the range.

Lovell Advocates Welcome Back the Reverend Floyd Schwieger

During the 1980s there had been little interaction between the BLM and Lovell wild horse advocates. Interest in the Pryor wild herd waned during this time, perhaps because the initial crisis had passed. In those years the primary source of interaction with the BLM was Russell Gaspar, attorney for the American Horse Protection Association, a national organization that filed a number of lawsuits against the BLM during this period. He stated in correspondence with Billings Resource Area manager Jerome Jack that he was also writing on behalf of Hope Ryden, who continued to visit the PMWHR during the 1980s and to maintain a count of the horses. This series of letters documented the gap between her numbers and the Billings Field Office census: the BLM's number was higher than Hope Ryden's count.[18]

A news release from Jerome Jack of the BLM Billings Field Office on March 26, 1982, stated that there were 150 adult head on the Pryor range and that an assessment of range conditions in 1981 concluded there was sufficient forage for only 120 adult horses. He noted that the last substantial capture had been in 1977. Although the harsh winter of 1977–78 had reduced the herd by almost half, their numbers had rebounded. The BLM proposed a three-phase capture plan to remove 25–30 horses: first by trying a water trap, second by trapping the horses using a fence, and finally, if the first two options failed, gathering them by rope from horseback. He estimated that all 25–30 horses would be adopted.[19]

In a letter to the Reverend Floyd Schwieger dated June 6, 1990, Hope Ryden welcomed him back to Lovell and expressed her concerns about what was happening on the PMWHR. She observed "two big grey horses (stallions) at the top of the mountain" and realized that the BLM was bringing in outside blood to the Pryor herd. She recalled in her letter, "All in all, the situation seemed too dismal and too overwhelming." The *Lovell Chronicle* publisher, Pat Schmidt, a strong supporter of the Pryor wild horses, had moved away, and when Hope Ryden gave a talk and slide show about the Pryor wild horses at the new Bighorn Canyon NRA Visitor Center, only five people attended. She was glad that the Floyd Schwieger had retired in Lovell.[20] This Lutheran minister would reenergize Lovell citizens and expand the base of wild horse advocates until his death in 2005.

By 1989 Floyd Schwieger had resurrected the PMWHA and recruited new members, including Mayor John Nickle, Lovell Chamber president David Peck, and Montana-Dakota Utilities manager Rick Hettinger. The group adopted a statement of purpose and series of goals and objectives, including the following: to support preservation of the Pryor Mountain herd physical and primitive-type characteristics; to support the Bureau of Land Management, the National Park Service, the US Forest Service, and other operating agencies in their efforts to improve the range for horses and wildlife; to promote public appreciation and awareness of the uniqueness of the Pryor Mountain range and horse herd; and to take a leadership role in establishing and operating a Pryor Mountain Western Wild Horse Historical Museum near Lovell.[21]

The original vision for the museum was a large, modern design with room for meetings and conferences, as well as a research library, permanent displays, and a gift shop. The current log-cabin structure maintains displays explaining the history of the PMWHR, the characteristics of the Pryor Mountain Mustang, and a gift shop with T-shirts, books, videos, and photographic artwork, all related to the Pryor wild horses. The PMWHA originally envisioned developing a coalition with other areas and groups where primitive-type wild horses were found, but in later years the board narrowed its focus to local issues. The scope of the original goals and objectives was broad, adding to its list of goals support for the BLM's idea of developing an overlook on the range with easy public access and appropriate area signs.[22]

The PMWHA supported the 1984 *Herd Management Area Plan,* in particular its objective to select for herd quality based on roan, mouse, and buckskin coloring and the small, compact horses found in the Pryor Mountains at that time. Under Floyd Schwieger's leadership, the new organization focused on the Spanish Mustang and feral or wild herds identified as having the unique colors and primitive-type markings associated with the Colonial Spanish horse. The PMWHA goals and objectives underscored the importance of managing the herd for the three most unique colors of Pryor horses: dun, grullo (mouse), and deep black, as well as primitive-type markings. The formation of a breeders association was a natural offshoot of members' interest in preserving and showcasing the Pryor Mountain Mustang.[23]

The PMWHA had its first meeting on September 6, 1990, at the Big Horn Restaurant in Lovell and elected five board members: the Reverend Floyd Schwieger, John Nickle, the Reverend Duane Duley, and Patti Herring of Lovell and Jim Edwards of Columbus, Montana. By this time Dr. Gus Cothran from the University of Kentucky had released the initial results of his genetic analysis, and the findings were momentous for many reasons. Blood sampling revealed a strong affinity between Pryor horses and Spanish breeds and no evidence of inbreeding. Even more significant, there was an absence of thoroughbred or draft-horse blood.[24]

What this study revealed was that the herd carried alleles linking the genetic makeup of the herd to the original Spanish mustangs roaming in the Pryor Mountain Complex:

> Preliminary results indicate that the feral horses have levels of genetic variability comparable to that of most breeds. In addition, we have observed genetic variants in the wild horses that we have not seen in any other breed of horse. Overall, the feral horses have the greatest genetic resemblance to the Spanish breeds for which we have data (three paso breeds and two South American breeds). Our study is only in its early stages, however the results to date indicate that some of the feral horses represent unique (from the standpoint of genetic markers) equine resources. As well, we see some support for the presumed "Spanish" ancestry of the wild horses.[25]

Lovell advocates were thrilled to discover that their efforts had led to establishment of the first public wild horse range and preservation of a unique strain of Colonial Spanish horse. The news bolstered the argument for marketing the wild horses as a tourist attraction, and Bighorn Canyon NRA superintendent Bill Binniwies informed the new organization that they were adding an extra dimension to visitor experience in the NRA by interpreting the natural history of the area for visitors.[26] The PMWHA envisioned building the first wild horse museum in the country and establishing a research center within the museum dedicated to the study of wild horses with national and worldwide significance. Their immediate concern, however, was an amendment to the BLM's 1984 *HMAP* that would reduce the size of the herd.

News of the change first surfaced at a fall 1990 meeting. The NRA superintendent explained why the NPS was withdrawing the Sorenson Extension from the PMWHR. The Park Service had issued two five-year special-use permits to the BLM, but monitoring of range conditions near Layout Creek in the northern (NPS) portion of the range showed that horses were making heavy use of the area and having a negative impact. The Park Service was also concerned about conflicting uses between wild horses and the nearly one hundred bighorn sheep. The BLM's *HMAP* noted that a small number of bighorn sheep occupied an area within the PMWHR along the Bighorn Canyon. In addition, the 1968 Secretarial Order creating a boundary that encompassed Park Service land had made it clear that the range was in all respects subordinate to the NRA so far as it affected lands within the NRA. The BLM could continue to negotiate with the NPS but would have to rebalance wild horse numbers in the event of a reduction. The current boundary reflects continued withdrawal of the Sorenson Extension.

In March 1991 Billings-area resource manager Bill McIlvain announced a decision to temporarily reduce the herd to eighty-five adult horses. He reported that all parties to a January 10 meeting had agreed to the plan, including the PMWHA, and the Reverend Floyd Schwieger would act as an equal partner in deciding which horses to release back to the range. The plan was to select horses displaying the strongest physical evidence of Colonial Spanish type. While news of the reduction was a blow to Lovell advocates, their response was consistent with past efforts to work with, not against, the BLM. For its part, the BLM needed "eyes on the ground" and familiarity with individual horses, as well as expertise in identifying Spanish characteristics. The agency relied on Floyd Schwieger, who spent many hours on the range naming the wild horses and taking countless photographs.

In his notes from the January 10 meeting, Schwieger observed that each agency came from a different philosophy and management approach. He also reported that because no one wanted to come forth with a definite proposal to address the conflict between the bighorn sheep and the Pryor wild herd, the PMWHA did. Its recommendation was to reduce wild horse numbers, remove all horses from the Sorenson Extension, and establish criteria for removing horses based on blood testing and keeping the best of the Spanish-type animals. Values for the PMWHA were that the proposal placed the horse program on a scientific basis, allowed some horses to remain visible to visitors along the road in the Bighorn Canyon NRA, and established the organization as making the first step toward discussion and compromise. As he added in his notes, "If there are any over grazing problems in the future, adjustments and relocations must be considered for the sheep. The horses have adjusted."[27]

Even before the Secretarial Order designating the PMWHR, Lovell wild horse advocates had avoided calling in outside help from humane organizations. The Humane Society of the United States had filed suit in 1968, seeking a temporary restraining order against the BLM, to prevent removal of the wild horses; however, only Lloyd Tillett joined the HSUS as a coplaintiff. The BLM had argued that it had no immediate plans to remove any wild horses. The federal district court judge then denied the request, but had left the complaint pending before the court so that plaintiffs could file an amended petition if the BLM acted to remove any wild horses. The lawsuit appeared to be a successful strategy, but PMWHA members had been working behind the scenes with the Billings Field Office and were aware of high-level meetings in Washington, DC, as recounted in chapter 1.

The PMWHA and HSUS once again took different approaches following announcement of a temporary herd reduction. While Floyd Schwieger was working with Billings officials to help them identify which horses to remove and release back to the range, the HSUS and the American Horse Protection Association (AHPA) joined several individuals who filed for an injunction against the BLM. The federal court judge in Billings found in favor of the BLM, but ordered the agency to justify its decision before any removed horses were adopted.[28] The removal proceeded in early April 1991, and Patti Herring of the PMWHA board spent the day at the Britton Springs corrals, helping the wranglers. She reported that the crew was taking care in bringing the horses in from the range and keeping the subherds separated. When one of the wranglers told her that a Kiger (introduced) mare had foaled in the corrals and needed help nursing her foal, she spent the day with the mare and foal.[29]

While the removal was proceeding, an attorney for the BLM argued before the federal judge that the agency was following the law by rebalancing wild horse numbers in light of the decision by the NPS to cancel the lease on the Sorenson Extension. The lawyer for the HSUS and AHPA, however, made a successful case that the *HMAP* in effect allowed for a reduction only to 120 adult horses. If the agency wanted to amend its plan, then there was a process to follow in reducing the maximum number of wild horses, including public hearings and an environmental analysis. The federal judge ordered that the herd population must be kept at 121 animals plus or minus 5 percent and that a specific number of horses be returned to each subherd area.[30] The number actually removed in 1991 was 16 adult horses, and the Billings Field Office began the process of amending the 1984 *HMAP* that fall.

In the midst of the legal controversy over the gather, an editorial appeared in the *Lovell Chronicle,* written by the publisher, David Peck, calling attention to the different philosophies and approaches of the PMWHA and outside organizations that had filed suit against the BLM:

> The roundup of the Pryor Mountain wild horses continues this week, despite efforts by some individuals and national groups to halt the roundup, and the issue brings to mind a few observations we feel need to be made. . . .
>
> The individuals who filed the lawsuit last week showed they mean business, and that's good. The members of the Pryor Mountain Wild Horse Association also are concerned about the well-being of the horses.
>
> We encourage all those concerned about the Pryor horses to work together. January's meeting of the minds among the various federal and state agencies was a good first step. And if the Humane Society and the American Horse Protection Association have money to file lawsuits, we urge them to spend some of that money locally for the betterment of the wild horses, for studies and support, rather than for paying lawyers.[31]

This editorial was pointed in its rebuke of outside activist groups. There would continue to be more lawsuits filed by other outside organizations; however, the "Lovell way" was to avoid getting directly involved in litigation and to seek ways to consult with the BLM and to work alongside outside wild horse groups. The message in this editorial was that the PMWHA, especially the Reverend Floyd Schwieger, was just as concerned about the wild horses as the HSUS and AHPA and that

Lovell advocates ought to be taking the lead. In the years to come, the PMWHA and its successor organization, the Pryor Mountain Wild Mustang Center, worked to ensure its niche as a local presence in public debates about the Pryor wild herd.

At the end of October 1991 the BLM held a public hearing on its proposed amendment to the 1984 *HMAP*. Three federal agencies had approved that document, the BLM, USFS, and NPS, because the herd management area encompassed lands within Custer National Forest and the Bighorn Canyon NRA, as well as within BLM public lands. The BLM was the lead agency for management of the wild horse herd; however, the withdrawal of the Sorenson Extension by the NPS drove home the point that the BLM had to balance its mandate to protect the Pryor wild herd with the Park Service's interest in having bighorn sheep. Tensions among the missions of the cooperating federal agencies became apparent at the public hearing.

The PMWHA, represented by Floyd Schwieger, recommended that if the BLM was going to reduce the herd size, it should retain the best of the striped primitive-type horses. When Bill McIlvain, manager of the Billings Resource Area, reminded people at the public hearing that the agency was charged with protecting the range for wildlife as well as wild horses, Schwieger rose to object that "I was here from the beginning and the horses must take precedence on the range. Deer and sheep are not unique. They are in abundance all over the country. But wild horses are not. This was established primarily as a wild horse habitat. The sheep are Johnny-come-lately's in the area."[32]

A year later the Pryor Mountain Wild Horse Association, renamed the Pryor Mountain Mustang Association (PMMA), solidified its consultative role with the BLM by developing criteria for future gathers and removals in response to interest expressed by Bill McIlvain in forming a "cull committee." Its concerns continued to focus on preserving the Colonial Spanish type in the herd known as the Pryor Mountain Mustang:

1. Cull all non-Pryor horses as soon as possible.
2. According to advice from Dr. Phil Sponenberg, try to maintain an equal number of horses for the seven basic colors (blacks, especially black skinned, blood, brown and mixed bays, white and

golden duns, dark and silver grullos, sorrels, blue roans and red roans.

3. Maintain the typical Pryor Primitive (Spanish) type found on the range when established. Do not try to upgrade these horses by the selection of larger horses as was done in the past. We are trying to preserve a type and not to make them adoptable. The original Pryor/Spanish type horse is a small/compact type horse. Most of these horses stand between 13.2 and 14.2 hands with an occasional larger horse. They should be kept so.
4. Maintain any unique new colors that may be developed in the Pryors such as Pinto, Palomino. Dr. Sponenberg feels this will happen.[33]

Lovell Advocates Seek to Preserve the Pryor Mountain Mustang

The 1992 amendment to the *HMAP* resulted in the removal of 46 adult horses, and the new population range of 85–115 animals raised concerns about the survival of the Pryor herd. The severe winter in 1977–78 had reduced the herd to 70 head, and Lovell advocates worried that another weather event could wipe out the entire population. Lovell advocates formed the Pryor Mountain Mustang Breeders Association in order both to ensure a genetic background herd should anything happen to destroy the present Pryor herd and to preserve and raise the best of the Spanish-type horses.[34] Elected board members were chair Art McGill, secretary Patti Herring, treasurer Bob Doerr, registrar Gail Good, and PMMA representative Floyd Schwieger.

Breeders could purchase their foundation stock from two sources, either wild horses adopted from the PMWHR or Tillett stock with a TX Ranch certificate. TX Ranch horses were descended from the Pryor wild horses that the BLM had allowed Lloyd Tillett to gather and move to his side of the boundary fence after the Secretarial Order creating the PMWHR. The PMMBA board required that anyone wanting to register their mustang had to show a BLM freeze brand or TX certificate, blood testing from the University of Kentucky, and four photos, showing front, back, and sides of the horse. Bylaws to the constitution stated that the purposes of the PMMBA were to cooperate with the BLM in protecting, promoting, and enhancing the wild herd; to work to establish "captive" herds to preserve a gene pool; to encourage use of Pryor Mountain Mustangs by their owners; and to promote activities among breeders and the interested public.[35]

Members of the PMMA shared many goals and members with the PMMBA. In January 1993 the PMMA newsletter assessed the fall 1992 roundup, noting that many of the finer horses ended up with members of both groups. A representative from the BLM proposed an advisory board to make recommendations about selection criteria and improvements in wild horse management. The PMMA, PMMBA, Lovell Chamber of Commerce, and BLM would each have one representative on the board. The breeders association turned to Dr. Phil Sponenberg from the Veterinary College in Blacksburg, Virginia, for help in developing a list of Pryor horses to remove and retain:

> The Pryor Mountain wild horse herd is the single most Spanish of the feral horse herds in the USA at this time. The history, phenotype and blood types of the herd all point to an origin from Spanish horses. This horse herd should be managed to maintain its uniqueness. This should include management to increase the excellent Spanish type already present, and should also maintain the color variation present in the herds. Management should also target numbers and sex ratios that will allow the horses to be maintained as an isolated population. Current lack of inbreeding in the population indicates that this is possible.[36]

In his 1993 evaluation Dr. Sponenberg made the point that, while there may have been opportunities for introduction of non-Spanish riding or draft-type horses by homesteaders, the mountainous terrain favored the survival of the smaller Spanish original. In his view removals of wild horses for use by ranchers and the US Cavalry would have favored the larger crossbred horses and left the smaller and truer Spanish type behind. He assessed both adopted and range horses and concluded, "The horses clearly converge on Spanish type, and some are remarkably Spanish in type." He cited blood-typing studies by Dr. Cothran, observing that the Pryor horses were closer than any other feral population being studied at that time, and he concluded, "The Pryor horses are a unique North American genetic resource, and should be managed to preserve that uniqueness."[37]

While Dr. Sponenberg emphasized the need to maintain sufficient numbers because the Pryor herd was an isolated population, he also favored removing horses introduced from the outside as well as horses that deviated from the Spanish type in order to avoid a mixed-breed herd. This assessment, including its recommended long-term strategy, formed the basis for advice to the BLM in subsequent removals: "Following the removal of these types the population would be of consistently top quality Spanish type horses. Removals could then shift to a long term strategy of removal of younger horses to maintain numbers, balance color variations, and other population goals such as balancing selection of replacements to assure that all families are represented." The PMMBA was also concerned about preserving Pryor Mountain Mustangs in the wild, as well as in "captive" herds.

In April 1994 Dr. Sponenberg again evaluated the horses and assigned them into four categories. The "super" horses had excellent conformation and old Pryor phenotype regardless of color and would produce more of the "super"-type horse. The "keep" category included horses with good conformation and rare colors, as well as young and very old horses that were more difficult to assess. The "average" category could be culled or retained, because the horses had no obvious flaws, nor did they have outstanding features. The "cull" category was made up of horses known, or suspected, to be introductions from outside the herd because their phenotype was so different from the old Pryor-type horse. When he shared the categories with Floyd Schwieger, PMMBA president Dale Hartman, and PMMBA registrar Gail Good, he was amazed by the unanimity of opinion on the "super" and "cull" horses. They ranked the horses without consulting one another, leading him to conclude that "there is indeed an old, traditional Pryor type horse recognizable by those with long experience specifically with the Pryor horses."[38]

There were fifty-one adult horses eventually removed during the 1994 gather, based in part on a spreadsheet for a horse inventory developed by Dr. Sponenberg in consultation with PMMBA representatives. In a postgather trip to the range, Dr. Sponenberg observed that all but one of the introduced horses had been removed and concluded that most of the horses remaining were now a very consistent Spanish type. He commended Dale Hartman, Floyd Schwieger, and BLM officials for taking a long-term approach to managing the Pryor Mountain Mustang and for their willingness to work together in identifying horses for removal and release back to the range. Finally, he recommended that BLM staff and Lovell-area volunteers continue to maintain an inventory and description of the horses.[39]

The number of registered Pryor Mountain Mustangs was growing, due to the leadership of Dale Hartman and Floyd Schwieger. A lively debate among active members from both the PMMA and the PMMBA arose over the question of what constituted pure Spanish traits. A "Roman" (convex) nose was an important criterion for some, while others focused on "line-back" duns with their zebra-striped legs. They were united, however, in their commitment to preserving this unique strain of Colonial Spanish horse. Their efforts to manage the wild herd for Spanish traits had quelled some of the earlier skepticism about whether the horses were actually remnants of the large "trespass" bands roaming in the Pryor Mountains at the turn of the century. Nevertheless, there were still some areas of controversy about the wild herd.

One issue was the long-term effects of introducing wild horses from outside herds. In addition to the Kiger horses from Burns, Oregon, a young buckskin stallion from the Rock Springs, Wyoming, herd management area was on the range until his removal in 1992. Although every effort was made to remove the introduced horses, several years had passed, and the extent to which those horses had already contributed to the gene pool was hard to determine, especially in the young horses. The appearance of a palomino filly on the range raised eyebrows among some Lovell advocates familiar with more typical Pryor Mountain Mustang colors. Nevertheless, BLM field staff decided to release the horse back to the range during the 1992 removal.[40] The filly was a cream-colored horse named Phoenix, whose palomino roan son, Cloud, became the most well-known horse on the PMWHR based on a series of PBS *Nature* films by Ginger Kathrens. Her documentaries helped to promote public awareness of the Pryor herd.

A second controversial issue was how BLM wranglers conducted the 1994 gather. A foal had been separated from its mother and died on the range. There were crowded conditions in the Britton Springs corrals at the base of East Pryor Mountain. An evaluation of the roundup by a committee of PMMBA and PMMA members recommended that wranglers release mares to find their foals if they became separated while being driven down the mountain. Less crowded conditions in the corrals might prevent aborted fetuses, and there should be more hay feeders. The consensus was that the system of identifying horses for removal worked well.[41] In spite of this evaluation by Lovell advocates, calls by some outside wild horse organizations for a review of methods used during the 1994 removal led to a BLM review panel and the appointment of a full-time wild horse specialist for the PMWHR.[42]

Lovell Advocates Continue to Favor Consultation over Conflict

Lovell advocates remained closely involved with BLM Billings officials following the controversial 1994 gather. One of the outcomes of the BLM review panel was that the new wild horse specialist was responsible for developing an implementation and monitoring plan, as well as action steps to reduce animal stress during future gathers. In February 1997 the PMMA met to discuss the draft implementation plan. Members noted that range conditions had been improving over a twenty-five-year period, exchanges among subherds promoted genetic diversity, and the average body-condition score of the horses was very good to excellent. The draft plan called for help from the public with photos of the horses, historical information about the range, and recruitment of volunteers.[43]

In September 1997 PMMA members reaffirmed that the organization's overall goals were to sustain a viable herd and to preserve the Spanish heritage of the wild horses. They discussed the upcoming gather and their desire to pick out the most typical Spanish horses to leave on the range. The models were Dale Hartman's stallion Dunbar and the Reverend Floyd Schwieger's favorite, a wild grullo mare with a two-tone mane and tail named Sacajawea.[44] A "super-horse" list developed for the 1997 removal included more than forty-five adult horses that should be released based on phenotype and rare color for the Pryors. Among the horses identified for their rare color were Isabella, a buckskin roan and mother of Phoenix; Phoenix, a palomino and mother of White Cloud (also known as Cloud); and Cloud, a two-year-old stallion identified as the most unique horse on the range. Other stallions and mares were listed as being excellent types.[45]

The PMMA was unique among wild horse organizations because its primary goal was the preservation of the Pryor Mountain Mustang. Activists belonging to national organizations monitored the welfare of wild horses across the West, often litigating actions taken by the BLM to remove and reduce wild horse populations. Lovell advocates, on the other hand, had a long-term relationship with BLM officials and a history of consulting with them on solutions to local management problems. They provided "eyes" on the Pryor wild horses and knew each by name, as well as by Spanish phenotype score. Members of the PMMA and PMMBA as well as BLM officials valued the history of the Pryor Mountain Mustang and agreed that preservation of this unique strain of Colonial Spanish horse within the wild herd was a management priority. Yet Lovell advocates also appreciated the "sense of place" associated with the PMWHR and were willing to support removals if scientific evidence demonstrated that wild horse overpopulation was destroying the fragile ecosystems on the Pryor range.

A Pryor Mountain Mustang foal named Jesse James torments his long-suffering mother.

This Pryor Mountain Mustang yearling shows no fear of humans. Young horses stay with their family bands until they are old enough to breed.

Pryor Mountain Mustang bachelor stallions spar with each other to perfect their fight skills.

TOP: "Cloud" is the star of three PBS *Nature* documentaries, filmed by the Cloud Foundation's executive director, Ginger Kathrens.
BOTTOM: Horses congregate around a water hole during the heat of summer, taking their turn according to an invisible pecking order among the family bands.

TOP: The unique features of the Pryor Mountain Mustang include small stature, short back, and a slight flare of the nose just above the nostrils. BOTTOM: A small family band approaches visitors near the road that runs through the National Park Service portion of the PMWHR.

A Pryor Mountain Mustang looks out over a ridge at the top of East Pryor Mountain.

A band stallion stands watch, ears alert, as his family grazes nearby, alert for predators and marauding bachelor stallions.

A Pryor Mountain Mustang romps in a pasture at Arrowhead Ranch in Baldwinville, Massachusetts. Owner Red Tolman is one of the few breeders still operating in the United States. (Courtesy of Red Tolman)

Chapter Three

Barn Raising

"Standing Up" the Pryor Mountain Wild Mustang Center

The Pryor Mountain Mustang Breeders Association was very active during the next decade, registering and showcasing Pryor Mountain Mustangs to the public. In addition, members of the Pryor Mountain Mustang Association made a key decision to build the Pryor Mountain Wild Mustang Center. Between 1998 and 2008 both organizations expanded their memberships and involved key donors from the Lovell area and from private foundations in the construction of the center building and in the development of educational programs about wild horses.

The Breeders Association Showcases the Pryor Mountain Mustang

By October 1998 there were 64 horses registered through the PMMBA. At the time of the original Secretarial Order in 1968, the gate to the boundary fence between the PMWHR and TX Ranch was left open for a period of time. According to the Reverend Kurt McNabb, who assisted Lloyd Tillett, the BLM allowed six months for the Tilletts to identify and claim horses belonging to TX Ranch. This reprieve gave Lloyd Tillett the right to move any horses he considered his own to TX Ranch pastures. Pryor Mountain Mustangs that he considered "his" were black, brown, and line-back duns.[1] Because Lloyd Tillett was a lifelong observer of the wild herd and familiar with the physical characteristics of Spanish-type horses, it is highly likely that he selected the best of the best. In fact, the PMMBA required that applications include either proof of adoption from the PMWHR or a certificate from TX Ranch.

The number of horses registered with the breeders association increased steadily over the next ten years, reaching 194 Pryor Mountain Mustangs in 2007.[2] These Spanish-type horses became a familiar feature at the annual Lovell Mustang Days parade, mustang horse shows, and horse expos. President Dale Hartman had a number of Pryor Mountain Mustangs, including his well-known stallion Dunbar, originally from the TX Ranch. Local members also included Dale Hartman's daughter and son-in-law, Cathie and Todd Nunn, and wife, Daphne Hartman. PMMBA members and breeders from outside Lovell included Bess Carnahan from Fort Laramie, Wyoming, on the Wyoming-Nebraska border, and Red Tolman and Arlene Dunham, from Arrowhead Ranch in Baldwinville, Massachusetts, who bought their first two mares from Dale Hartman, a stud colt from Cathy Nunn, and a mare from Floyd Schwieger.

This period was, in many ways, the "golden age" of the Pryor Mountain Mustang. The members of the PMMA and PMMBA supported one another's organizations, as well as annual public events such as the parade and local horse shows. Owners of registered mustangs attended the 2001 BLM adoption event to demonstrate that Pryor wild horses were highly trainable. During the planning for and construction of the new center, local members of the breeders association brought their horses for "Meet the Mustangs," where Lovell residents could see and touch them. There was a new level of awareness and excitement about Pryor Mountain Mustangs inspired by Dale Hartman and the Reverend Floyd Schwieger that resulted in widespread support for the new center, including financial contributions from local banks and businesses as well as private foundations. Still, the number of breeding-age horses in the registry was small, and a catastrophic event on the range could have wiped out most of the gene pool.

Lovell Advocates Plan the Pryor Mountain Wild Mustang Center

By 1998 members of the PMMA decided that there needed to be a research center or museum focusing on the establishment of the PMWHR and studies of the social structure and behavior of wild horses. A great deal of excitement accompanied the initial planning sessions for the Pryor Mountain Wild Horse Center, Inc. (PMWHC),[3] as reflected in the bylaws: "The purposes of the Pryor Mountain Wild Horse Center, Inc. are as follows: To build and operate a facility to serve as an educational institution for the Pryor Mountain Wild Horse Range, a research center for the history of the wild horse, and a training center for wild horse management and the general public."[4]

Early drafts of promotional literature established the importance of the new center as providing a sense of place tied to the PMWHR, as well as a unique identity for Lovell and an attraction for the public. These aspects of the center's vision reflected the aspirations of the earliest Lovell advocates who made the case in the 1960s that the wild horse range would help to grow the tourist base and stimulate local economic development.[5]

During the 1960s the Lovell Chamber was the focal point of efforts by members of the historical committee to save the Pryor wild herd. "Standing up" a new center, however, was a frontier-style barn-raising event, requiring manual labor and financial contributions from the Lovell community. The town council agreed to sponsor a ten-thousand-dollar matching grant from the Wyoming Business Council for a planning project, while local banks, including the Bank of Lovell, First National Bank Powell (Lovell branch), and Big Horn Federal Savings and Loan, and several board members donated money toward the fifty-fifty local match.[6]

The planning grant funded a retreat in November 1999 attended by PMWHC board members and Linda Coates-Markle, the BLM wild horse specialist, and facilitated by Plan One/Architects. The plan envisioned a building with a lobby and exhibit space, a gift shop, an orientation theater, a conference room, classrooms, and staff offices. The back would open up to a corral and pasture with wild horses. The potential site for the center was west of the National Park Service Visitor Center, located on Route 14A that connected with Highway 37, the road through the Bighorn Canyon National Recreation Area.[7] The center would be the first wild horse museum in the country.

The core group of Lovell advocates involved in starting the new center was a small but committed network of wild horse lovers, some with ties to the original campaign to create the wild horse range in the 1960s. The first president of the PMWHC board, John Nickle, moved to

Lovell as town manager during those early years and became involved because of his friendship with the Reverend Floyd Schwieger. At that time Cal Taggart was mayor of Lovell; he and his wife, Irene, became strong supporters of the new center, pledging fifty thousand dollars of their Charitable Remainder Unit Trust with a matching condition to the PMWMC in 2002.[8] One of the founding members of the PMMA in the 1960s, Jim Edwards, a real estate developer and mustang breeder from Fishtail, Montana, continued his involvement as a member on the center's board of directors. The board's secretary, Daphne Hartman, was actively involved with the breeders association.[9] A common link was to Floyd Schwieger, who drew in new board members and sustained relationships with the earliest Lovell advocates.

Connections between members of the PMWHC board, Lovell town officials, and the banking community were instrumental in getting plans for the new center off the ground. The "sense of place" reflected in the promotional literature was evident in widespread support for the initial planning grant. Lovell advocates had for years identified with the Pryor wild herd as being part of their community; now their dream of having a physical structure to anchor that sense of place was to become a reality. The proposed location for the building, west of the NRA Visitor Center, had a direct line of sight to East Pryor Mountain. Travelers passing over the Big Horn Mountains on their way to Yellowstone National Park could pause for information about the Pryor wild herd and for driving directions to the Dryhead herd area within the NRA or to Burnt Timber (Tillett Ridge) Road and the challenging two-hour trip to the meadow on the top of East Pryor Mountain, where most of the Pryor wild horses spent the summer and early fall.

The next crucial steps were to purchase the land proposed for the new center and to design a logo and fund-raising brochure. The Town of Lovell approved thirty-five hundred dollars for a logo design and for printing fund-raising brochures, while the Lodging Tax Committee approved three thousand dollars for the brochure design.[10] This kind of public support reflected the importance that citizens of Lovell and their elected officials attached to the new center. The Pryor wild herd had become a key part of the town's identity, and the prospect of enhanced tourism dollars for the local economy motivated local government investment in the project. Equally important for the citizens of Lovell was the educational mission of the new center and the potential benefit for students to learn more about the role of wild horses as a part of their heritage. The board enjoyed widespread enthusiasm, but its members still had a lot of work ahead of them.

Fortunately, John Nickle was the former town manager as well as the former mayor of Lovell. His history with establishment of the PMWHR, his knowledge about the Pryor Mountain Mustangs, and his professional training were all invaluable in applying for numerous grants to purchase the land and construct the building. He spent countless hours pursuing foundation and government funding, in addition to making personal donations. The PacifiCorp Foundation donated substantial sums to the center. After the center was open, individual contributions from visitors and supporters who followed the wild herd via the center's website blog helped to sustain day-to-day operations. In January 2001 the board voted to rename the nonprofit corporation the Pryor Mountain Wild Mustang Center and completed purchase of twelve acres next to the NRA Visitor Center with a loan from the Town of Lovell.[11]

An editorial by publisher David Peck in the *Lovell Chronicle* pointed with pride to "strong signs in recent years that we are a community that wants to move in the right direction, and one of the projects leading the way is the proposed Pryor Mountain Wild Mustang Center." He urged the citizens of Lovell to embrace the proposed center: "As the gateway to the first public wild horse range in the United States with the best examples of Spanish mustangs known to exist, Lovell is a natural place to host a wild mustang center of national prominence and scope. Those who are working on this project intend the wild mustang center to be the quality of the Buffalo Bill Historical Center but on a smaller scale. In fact the board of directors has been consulting with officials at the Buffalo Bill."[12]

The list of contributors at this stage of the project showed the extent of support from local businesses and organizations, as well as individuals, many of whom gave multiple times: the Pryor Mountain Breeders Association, Pryor Mountain Mustang Association, Bank of Lovell, First National Bank, Big Horn Federal Savings and Loan, North

Big Horn Development Council, Town of Lovell, Big Horn Mt. Medicine Wheel Association, Big Horn Engraving, and Montana-Dakota Utilities Resources Foundation.[13]

The composition of the PMWMC board included individuals who had made personal contributions over the years and those who had contacts with the business community. Board president John Nickle owned Lovell Cable TV and was the former mayor of Lovell. Another former mayor, Cal Taggart, also served on the board, as did town manager Todd Wacaser. The Lovell volunteer fire chief James Minchow was a local petroleum-products distributor, and Jim Edwards, a real estate developer and mustang breeder, was one of the two original PMMA members still on the board, along with Floyd Schwieger. Board secretary Daphne Hartman was chief financial officer of North Big Horn Hospital District. The center board also included two outside advisers, Hope Ryden Miller and veterinarian Dr. Phil Sponenberg.[14]

Lovell Volunteers Carry on the Legacy of Floyd Schwieger

Business contacts and fund-raising with large donor prospects were critical to the goal of building the Pryor Mountain Wild Mustang Center. Several foundations rejected applications to fund construction, but John Nickle reported to the board in early 2005 that they had received sixty thousand dollars plus six thousand dollars in pledges from Georgia-Pacific Corporation, Montana Dakota Utilities, and the Bighorn Canyon NRA toward the estimated cost of eighty-five thousand dollars to build a log structure. The First National Bank offered an interest-free loan for any balance. The board agreed to proceed with the construction project.[15] The groundbreaking ceremony took place in late March 2005, officiated by Big Horn County commissioner Keith Grant, Lovell mayor Bruce Morrison, Georgia-Pacific plant manager Bruce Wacker, and the Reverend Floyd Schwieger, who passed away within four months of witnessing the start of his dream for a center.[16]

The death of Floyd Schwieger on July 20, 2005, was a blow to Lovell advocates, especially those who had worked with him during the nearly forty years it took to establish the PMWHR, consult with the BLM to preserve the Spanish bloodlines in the wild herd, establish a breeders association to preserve a Pryor Mountain Mustang gene pool, and plan for the center. As the remembrance in the fall 2005 newsletter put it, "He found the Pryor horses, or they found him, in the mid-1960s. Rev. Schwieger was part of the original effort to save the herd and worked for their welfare the rest of his life."[17] An earlier news article about efforts to save the Pryor wild horses wrote that "the Rev. Floyd Schwieger comes to the table armed. Not with gun or club, to be sure, but with weapons appropriate for a 76-year-old retired Lutheran minister ready to defend the doctrine."[18]

Those who survived him often explain why they became involved with the Pryor wild horses by recounting how Schwieger inspired and recruited them to serve on the PMMA and then the center's board. He left behind close friends, such as Dale Hartman, with whom he had started the Pryor Mountain Mustang Breeders Association. Newcomers, as well as veterans, on the center board speak of carrying on the legacy of Floyd Schwieger, sometimes with tears in their eyes. He left a spiritual as well as financial inheritance: a fifty-thousand-dollar bequest to establish an endowment fund in his name.[19] After his death board members continued work on building the new center, but also persisted in raising issues first placed on the BLM's agenda by Schwieger, including range expansion. Educating the young about the history of the Pryor wild horses was another passion, and before his death he recruited to the center's board Nancy Cerroni, curriculum coordinator for the Big Horn County School District No. 2.

Individuals and businesses from the Lovell community stepped up to prepare the site for construction. As with a frontier-style "barn raising," neighbors donated time and equipment in the fall of 2005 so that construction could begin the next spring. Local companies included TCT West, G-K Construction, Frost Construction, and Lovell Concrete.[20] Montana-Dakota Utilities provided natural gas service installation, and Pacific Power and Light provided electrical service installation at no cost. The Town of Lovell provided water and sewer lines plus equipment and labor.[21] By the fall of 2006 the log walls were up and the roof almost finished. The fall 2006 newsletter reported, "5 J Contracting of Lovell and a host of local area volunteers have worked together providing the many hours of precise hand labor required for a good log structure."[22] By the time the grounds had been planted and seeded and the building completed, more than forty area residents had given an estimated twenty-eight hundred hours of labor and special skills.[23]

Even before the center's opening in 2007, visitors could see Pryor Mountain Mustangs owned by Dale and Daphne Hartman in the large pasture that was part of the center's land. A sign next to the pasture informed visitors that these "Horses are Pryor Mountain Offspring." They could see the unique Spanish characteristics of these mustangs—their dun and grullo colors and primitive line-back and leg zebra stripes. The breeders association was a key partner with the PMWMC, giving people a "hands-on" experience. "Meet the Mustangs" was held for the first time on the site of the new center in July 2004, attended by more than 175 guests from around northern Wyoming and twelve other states. Visitors were able to pet the Hartmans' horses and learn more about the plans for the new center. One of the stars of the event was a mare named Little Orphan Annie and her 2004 foal, El Reno. The mare was one of rare twins born on the range in the spring of 1994 and assumed to have died, but she surprised everyone by showing up during the 1994 gather, even though she was in very poor condition. The Hartmans adopted her and showed her at the National Wild Horse and Burro Show in Reno, Nevada.[24]

Meet the Mustangs was a popular annual event, educating Lovell-area citizens and visitors about the importance of preserving this unique strain of Colonial Spanish horse. New board member Nancy Cerroni presented slides showing Pryor family bands on the range and explaining about social structure and behavior based on her countless hours of observation and photography during the summers.[25] The educational mission of the center was especially important to her and was a major reason the Reverend Floyd Schwieger recruited her to join the board. She learned the names of each wild horse by showing her photographs to him when he was no longer able to make the trip to the range.

As a result of this extensive fieldwork, she developed an elementary school curriculum module distributed to teachers around the country. Her son, Matt Dillon, accompanied her on summer trips to monitor the wild herd and earned a college degree in environmental science. This education provided a unique perspective on restoring rangeland ecosystems for the health of the wild herd. He became the center's first director in 2007, and he developed an active working relationship with the BLM's new wild horse specialist, Jared Bybee. As the fortieth anniversary of the Secretarial Order approached, the long-term consultation between the BLM and Lovell advocates continued to evolve, although one issue stymied efforts to expand the PMWHR: the boundary with Custer National Forest.

Lovell Advocates Seek Expansion of the PMWHR

Expanding the range boundary was a crucial step toward the center's goal of increasing the size of the wild herd. The 1992 amendment to the *Herd Management Area Plan* established as an "Appropriate Management Level" (AML) 85–115 adult horses, but this number was likely too small to ensure genetic variation and to prevent inbreeding within the wild herd. There had been some additional acreage to the range after passage of the Wild Free Roaming Horses and Burros Act in 1971, but those modest expansions were based on studies of where the Pryor wild horses were found at the time of the law's enactment.[26] The 1973 Hall study had recommended adding a portion of the Lost Water Canyon area in the national forest; however, Floyd Schwieger, Hope Ryden, and others who had spent time with the wild herd during the 1960s insisted that the Pryor wild horses made historical use of other forestlands and that those areas should be included within the boundaries of the PMWHR.

In a 2000 letter to Sandy Brooks, the BLM Billings Resource Area manager, Floyd Schwieger recollected that "there were scattered numbers of wild horses on East Pryor Mt. Forest Service lands when the 1971 Act was passed. This was historic wild horse range. . . . My own visual observations have confirmed this."[27] He followed up with a letter to Secretary of Agriculture Ann Venamen in which he identified one potential addition known as the Dryhead Overlook, an old Forest Service grazing allotment at the top of the mountain overlooking the Bighorn Canyon:

> While there are still thousands of wild horses in the West, there are only a little more than 500 of these original Spanish Colonial type horses in

> existence today in five small bands. These are the type of horses first brought to our country by the Spaniards and once roamed the west in the millions.
>
> In 1968 America's first "Wild Horse Range" was established by Sec. of Interior Stewart Udall on the Pryor Mts. However, a good portion of their historic range was never included at that time. The area we are speaking about is known as the Dryhead Overlook allotment of the Custer National Forest.
>
> We have met with and spoken to the representatives of the Custer National Forest on a number of occasions. They have for various non-plausible reasons, at least to us, seen fit to deny our request. We know there were horses on the Dryhead allotment in 1971 and so should be allowed and managed there. They did indicate, however, that if we wanted this area added the will of the people must be made known. Hence we are engaged in this effort.[28]

Based on subsequent correspondence to Secretary Venamen, Floyd Schwieger received no acknowledgment from her. Shortly thereafter, John Nickle wrote to her about plans by the US Forest Service to construct a new metal fence replacing the buck-and-pole fence along the northwest boundary of the PMWHR. Wild horses were moving through the old boundary fence back into the national forest, and this letter requested continued use of the area by the horses.[29] In an ironic reference to the past, these Pryor horses were now in trespass once again and needed to stay on the PMWHR side of the boundary fence or possibly be removed during a BLM gather. The questions of what to do with the bands on forestlands, the boundary fence, and possible expansion of the PMWHR became a major issue for the board members of the new center.

Before he died Floyd Schwieger met with Ginger Kathrens, the filmmaker who had produced two PBS *Nature* documentaries that featured the life of the palomino roan stallion Cloud and his family band. Her purpose for the meeting was to record his observations about where the wild horses were found in 1968 at the time of the Secretarial Order. She recorded his recollections about the presence of wild horses near the Dryhead Overlook in areas known as Tony Island and Big Ice Cave in Custer National Forest and had a notary present to swear him in as an expert witness. This testimony was intended as evidence in a lawsuit to be filed by the Cloud Foundation against the US Forest Service in an effort to move the boundary line farther to the north and west. The center's board discussed joining the lawsuit as a coplaintiff, but members decided that there needed to be good scientific data on the location of the wild horses before they could be involved. The litigation was ongoing for a number of years, but the center's board never joined the lawsuit.[30]

Range expansion continued to be an important topic of discussion in 2006, and adding land within Custer National Forest was a key objective. The issue had been important to the PMMA before the center was incorporated, and John Nickle remembered a meeting in the 1990s with the wild horse specialist Linda Coates-Markle, in which she expressed the opinion that the wild herd ought to be maintained at approximately 150 breeding head or 200 adult horses. He resolved to find notes from that meeting and contact her about the center being able to attend BLM staff meetings in order to know more about range policy.[31] The center's board carried on the "Lovell way" established by the earliest advocates and tried to use their contacts and positive relationships with USFS officials to accomplish what Ginger Kathrens was trying to achieve through litigation.

While board members had reason to be optimistic, based on an earlier meeting with deputy regional forest director Kathy McAllister, they began to sense that the USFS had serious reservations about allowing wild horses onto forestlands. In particular, there were environmental reasons, including discovery of a plant that might qualify for listing under the Endangered Species Act.[32] Lovell advocates believed small bands could coexist with the flora and fauna of the forest, but forestry officials viewed wild horses as a species of large grazer that could damage fragile ecosystems.

A sense of urgency about range expansion was evident in John Nickle's board update in February 2008. He reported on an upcoming meeting scheduled at the center with Steve Williams, Custer National Forest supervisor, about possible expansion of the range farther into the forest. He urged board members to attend. Another meeting was coming up with the new BLM wild horse specialist, Jared Bybee, who wanted to update the board on the year's plans for the range and horses. There were other meetings or correspondence in progress with regional and national officers of the BLM and Forest Service, as well as congressional delegations from Montana and Wyoming. The president's update ended with this pointed message: "I believe we need to provide the institutional leadership in showing our friends and supporters how to help save the horses 'again' 40 years later."[33]

Several board members attended the meeting with Steve Williams on February 21, 2008. In his meeting report John Nickle wrote:

> We told Steve of the Horse Range history and he was very interested in learning about the horses. We showed him our Range expansion proposal and told him about many other Range and horse issues. He listened intently and asked many questions. He will research the 1968 Range establishment and how the expansions were done, and will also research the 1971 Act, to determine how it affects the Forest Service lands. One of his comments was "are you sure the proposed additions will provide enough more land to ensure a genetically viable herd?" He appears to have an open mind at this time and seemed to want to learn everything he could about the horses and the Pryor Mountains. It is a good start.[34]

The center's range-expansion proposal referenced in the letter included 4,412 acres in the area known as Tony Island, in addition to the Sorenson Extension on National Park Service lands in the Bighorn Canyon National Recreation Area and 600 acres of private land in the Tillett Trust. The total acreage in the expansion proposal was 14,000 acres. In an effort to move the proposal forward, John Nickle wrote to Hope Ryden, asking for photos of wild horses she had taken in 1970 on Tony Island, as well as a certified statement about where and when they were taken. He added that the center would use them, along with statements gathered from others who were involved in the first Pryor wild horse roundup in 1970. Attached to his handwritten note were typed minutes of center board meetings with Steve Williams and Mary Heller, field representative for Montana congressman Denny Rehberg.[35]

A separate center board report summarized a meeting with the new BLM wild horse specialist, Jared Bybee, on February 26, 2008.

The board learned that the BLM had yet to receive funding for a new *Herd Management Area Plan,* but the Billings Field Office was planning to remove thirty, mostly young, horses using a water-trap method. Instead of driving or roping the horses, they would trap horses in small pens near water sources, select the ones they wanted to remove, and then transport them to corrals at the administrative pastures known as Britton Springs. The current (1992) *HMAP* provided the authority for this removal. At the end of the meeting they discussed the possibility of a Memorandum of Understanding between the BLM and the center so the PMWMC could officially support the BLM with frequent monitoring visits, family lineage records, and a separate agreement for volunteer support of range-improvement projects that Jared Bybee hoped to include in a new *HMAP*.[36]

A number of interrelated issues were on the board's agenda, now that the center log-cabin building was completed and a director was hired. Range expansion was the top priority, because the permit to use the Sorenson Extension in the Bighorn Canyon NRA was likely to be canceled permanently. The remaining land at the base of East Pryor Mountain received very little precipitation, and forage was sparse. The situation at the top of the mountain was better in the sense that the subalpine meadows could support the wild herd during the summer months; however, the soils were fragile, and the grasses could support only a limited number of horses. While Lovell advocates had always understood the importance of periodic removals, they were worried about the potential for inbreeding within the small herd of one hundred adult horses. Preserving the Pryor Mountain Mustang was their ultimate goal, and without range expansion there was little hope for an effective long-term management plan.

The Center's Mission Becomes Established

The center needed to fund its operations once the log building was finished. Visitors stopped to study the exhibits, watch a video about the PMWHR on the touch-screen computer kiosk, and browse through the gift shop and library. The new director, Matt Dillon, made weekly trips to different parts of the range, recording information on the changes in band composition, location, and general condition. In order to sustain these operations, the center launched the "Wind Drinker 1000" fund-raising campaign in 2006. The name reflected the impression that the wild horses seemed to be drinking the wind as they galloped across the range. A photograph taken by Nancy Cerroni of two stallions streaking across a meadow of lupine became the symbol for the fund-raising campaign.[37]

The goal of Wind Drinker 1000 was to receive one-hundred-dollar gifts from one thousand donors. Name plates on the center's west wall under Nancy Cerroni's large framed photograph identified the donors. Once the center reached the goal of one thousand donations, donor names would be entered in a drawing to win a limited-edition bronze sculpture of a three-horse Pryor Mountain Mustang family.[38] Nineteen other small sculptures were offered to the public for a donation of ten thousand dollars each. The two-thousand-dollar difference between the cost for the center to purchase each small sculpture and the amount raised from each individual donation would go toward a life-size bronze of the same sculpture to be set on the grounds of the center. The artist selected for the project by the board was T. D. Kelsey, who was internationally renowned and locally recognized because of a one-man show at the Buffalo Bill Cody Museum.[39]

The center had benefited from the financial and in-kind support of local banks and businesses, as well as individuals from the community during the "barn-raising" phase of the center's beginning. By 2008 forty years had passed since creation of the PMWHR, and the center's board and staff were starting to establish goals for the future. The center's greatest challenge now was to sustain day-to-day operations in order to fulfill its mission, to preserve the Pryor Mountain Mustang. During 2008 the center welcomed more than two thousand visitors from forty-nine states and more than twenty foreign countries. Many signed the guest book and left donations. Their names were added to the center's mailing list and became potential future supporters of the Pryor wild herd. A small but faithful number followed Matt Dillon's blog that included photos and reports about individual horses and family bands.

A genealogy of the Pryor wild herd was a major joint venture for the center. The BLM Billings Field Office shared data from 1996 to

2006, as well as notebooks, labeled photos, and an identification system with the mother of each horse covering the years 1976–90. Those early records had been kept by BLM wrangler Lynne Taylor and were thought to have been lost, but were discovered in the Billings Field Office archives. The years 1990 and 1994 were covered by field notes and photos taken by Floyd Schwieger and lists developed by Dr. Sponenberg. Beginning in 2006 Matt Dillon tracked the horses and compiled records into a database. The genealogy project became a critical component of herd management, allowing the BLM to base gathers on preserving separate "kinship" lines.[40]

The board unanimously approved a Memorandum of Understanding with the BLM in 2008 that would authorize the center to help with herd monitoring once a new *HMAP* was finalized. The protocol stated:

> The BLM and PMWMC share many common goals for the management of wild horses. Both entities actively promote healthy, self-sustaining wild horses on public lands for future generations. The BLM and PMWMC also recognize that the interest of the protection of wild horses consistent with Public Law 92-195 is paramount to the mission of both entities. In an effort to educate the public about wild horses from the PMWHR, other wild horses from public lands in general, and the Wild Horse and Burro Program, the following will be agreed to:
>
> The BLM will share data related to management of the Pryor Mountain Wild Horse Range and Pryor Mountain Wild Horses; direct the public at large to the PMWMC for information related to the wild horses of the Pryor Mountains.
>
> The Pryor Mountain Wild Mustang Center will recognize the management of the PMWHR, wild horses and public interest the BLM Billings Field Office provides; develop and share wild horse demographics lists, tables, references and etc., such as foal lists, bands, family groups and other pertinent demographic data.[41]

Although the *Herd Management Area Plan* and authorization of projects under the Memorandum of Understanding were delayed by an appeal of the *HMAP* from the Cloud Foundation, the agreement between the BLM and the center reflected the extent to which the Billings Field Office had come to rely on the center. Members of the

PMMA, the Reverend Floyd Schwieger in particular, had historically helped with monitoring of the wild herd and consultation during periodic removals and adoptions. The scientific expertise of the new center director, however, proved to be an invaluable resource for the new BLM wild horse specialist who had a special interest in developing objective criteria for selective removals and fertility control, as well as plans to stabilize forage and water conditions. The *Herd Management Area Plan* called for an increase in the "Appropriate Management Level" from the 85–115 range that had existed since the 1992 amendment to a new range of 90–120 adult horses. This increase would be possible because of improvements on the range that would support a higher number of horses.

Planned range-improvement projects included adding water "guzzlers" to catch and store rainwater located on the midslopes of East Pryor Mountain. The idea was to encourage bands to spend more time in these areas where forage was more plentiful. The horses were accustomed to moving from the lower elevations to the subalpine meadows as soon as the snow began to melt, and the hope was that the presence of water sources in the midslope areas would reduce grazing pressure at the top of the mountain. If the horses would move more slowly up the mountain and reach the meadows in midsummer, then the rate of forage utilization would drop to more acceptable levels. In the long run, range improvements could increase the carrying capacity of the area and raise the Appropriate Management Level.

The Memorandum of Understanding became the target of harsh criticism by several wild horse organizations, because the center was—in the eyes of many wild horse activists—colluding with the BLM's long-term agenda of driving all wild horses from public rangelands. The PMWHR was a particular target of criticism because of the popularity of the "Cloud" PBS documentaries. The Cloud Foundation posted many news releases to its members and to the general public during the next several years, as the BLM conducted a large removal and started a systematic fertility-control program on the PMWHR. Relations between the Cloud Foundation and the PMWMC became strained during this time, especially in 2009 during the gather, when Matt Dillon worked on site as a volunteer to help wranglers identify and keep together family bands being driven by helicopter to holding corrals.

Lovell advocates had long understood that a catastrophic weather event or sustained conditions on the range, such as prolonged drought or a thick ice glaze on top of scarce forage, might wipe out the entire herd. They viewed the horses as wild, but realized that the boundary lines around the range had created an artificial environment and therefore a need for humans to care for the habitat on which wild horses depended. The center had consistently taken the position that "nature" should not regulate the wild horse population. Mountain lions were, in theory, a "natural" predator, but the prey was usually newborn foals, and it was unlikely that predation would stabilize the population.[42]

The center's mission continued to state its purpose as "dedicated to preserving and interpreting the Pryor Mountain mustangs, their evolution, history, habitat needs and historical significance." Its vision statement echoed the vision of the earliest Lovell advocates: "Preserve and promote a genetically viable herd of Colonial Spanish American horses in the Pryor Mountains."[43] An alignment with the BLM's plan for managing the Pryor wild herd was evident in the 2009 *Herd Management Area Plan:* "The population would be managed for a phenotype reminiscent of a 'Colonial Spanish Mustang' as described by Sponenberg. The population would be managed in a manner designed to preserve genetic traits, bloodlines and ensure maximum genetic variation within a small population while managing for healthy rangelands. The wild horses would be managed for an even sex ratio as well as age classes. Emphasis would be placed on retention and increasing the number of 5–10 year animals as a core breeding population."[44]

The center board also took a position in support of managing the Pryor wild herd using temporary fertility control, or PZP, as a strategy for balancing the numbers of horses with the carrying capacity of the PMWHR, while reducing or eliminating the needs for large gathers. A position in favor of temporary fertility control was opposed by the Cloud Foundation at that time and added to the tension between the two organizations. This difference in philosophies regarding artificial versus "natural" on-the-range management persisted for several years.

By 2012 TCF took a position in favor of aggressive use of PZP to eliminate the need for future gathers because of a concern that demand for adoptions was declining.

Opposition by the Cloud Foundation and other wild horse organizations to the positions taken by the center board occurred in the midst of major challenges to the management of the wild horse program at the national level. The BLM planned large removals across the western states to bring wild herds to their AML, as required by amendments to the 1971 law. The elemental forces of a "perfect storm" were brewing, as plans for large-scale removals in nearly every herd management area coincided with a major drop in private adoptions of captured wild horses and an increase in the number of wild horses in costly short- and long-term holding facilities. The center's board agreed that the BLM needed to develop and implement modern policies and management strategies, including the use of temporary fertility control. The PMWHR would prove to be a model of twenty-first-century management, as the BLM worked together with the center to implement small-scale gathers and fertility control that were tiered to the 2009 *Herd Management Area Plan.*

Chapter Four

A Perfect Storm

Engaging the National Wild Horse Debate

A "perfect storm" was developing in 2008, combining large-scale wild horse removals by the BLM across herd management areas in western states with reduced demand for adoptions and rising costs for BLM short- and long-term holding facilities. When the Billings Field Office drafted a new *Herd Management Area Plan,* the center found itself in an increasingly turbulent political environment. Lovell advocates had consistently chosen negotiation and consultation over litigation to preserve the Pryor Mountain Mustang during the past forty years, and they had declined to join the Cloud Foundation in its lawsuit against the BLM. Now the center was itself the target of criticism by outside wild horse activists for working with the BLM.

The 2009 *HMAP* Solidifies Preservation of the Pryor Herd

The original *HMAP* in 1984 had established the appropriate size of the Pryor wild horse population as 121 adult horses; however, the 1992 amendment had reduced the upper limit to 105 animals. Lovell advocates and outside wild horse activists shared a concern about the risks posed to the herd by this reduction: the potential loss of genetic diversity over the long term as well as the possible loss of the entire herd to a catastrophic weather event. Yet the facts were that the range boundaries were now constricted by the cancellation of the Sorenson Extension in the Bighorn Canyon NRA, as well as the potential for permanent closure to horses of areas within Custer National Forest outside the boundary of the range. Moreover, fragile ecosystems were susceptible to overgrazing, a problem exacerbated by prolonged drought.

An evaluation of range conditions in 2008 concluded that the "hands-off" management practices of previous years had resulted in deteriorating range conditions. The proposed HMAP increased the population's upper limit from 105 to 120 in part because it included proactive steps to prevent further range deterioration: development of water sources, especially in the midslope areas to encourage a wider distribution of horses, and treatment of noxious weeds that were invading and replacing nutritious forage plants. The proposed plan provided for periodic removals as well as fertility control, both controversial management practices among wild horse activists. These actions, however, made it possible to increase the population limit by taking steps to stabilize conditions on the range.

The 2009 proposed HMAP also identified objectives for maintaining herd characteristics based on the phenotype score of Colonial Spanish–type horses, including unusual colors; a sex ratio balanced evenly between males and females; a core breeding population of five- to ten-year-olds; and managing to ensure that each active breeding mare would have at least one offspring to carry forward to the next generation.[1] Phenotype scores and bloodline representation were objective criteria for managing the population, compared to the 1984 HMAP as amended. Of course, members of the Pryor Mountain Mustang Association and Pryor Mountain Mustang Breeders Association had been evaluating the herd and making specific recommendations to the BLM since the early 1990s; however, references in the previous HMAP were very general, stating as an objective, "Maintain the various colors within the herd and retain those horses with better conformation so as to gradually improve the quality of horses."[2]

Preventing the extinction of rare bloodlines in the Pryor wild horses was essential to ensuring their long-term survival, especially because of the limited numbers in the herd that could be supported within the boundaries of the PMWHR. If the BLM planned to implement PZP, or temporary fertility control, to limit further population growth, then it would be very important to allow each mare to have at least one offspring on the range. The kinship method of removing young horses whose bloodlines were already well represented on the PMWHR resembled the practice on Assateague Island National Seashore to limit the growth of a wild herd that had lived there for centuries.[3]

The Billings Field Office manager, Jim Sparks, issued a "Finding of No Significant Impact" based on the Environmental Assessment (EA) accompanying the *HMAP* and a "Decision Record" approving the plan in May 2009. The Cloud Foundation exercised its right to appeal and also to petition for a stay of the decision to the Interior Board of Land Appeals in the Office of the Secretary of the Interior. This petition had the effect of "freezing" the new *HMAP*, with its range-improvement projects and proposal to increase the Appropriate Management Level, keeping the current AML in place until the administrative judge rendered a decision. In a November 3, 2009, order that denied the petition for a stay, the administrative judge noted, "In addressing the immediate and irreparable harm criterion, appellants stated only that 'BLM may soon undertake removals of a large number of horses based on the challenged *HMAP*' and that the removals threaten the continued existence of the herd."[4]

The administrative judge reasoned that the decision to approve the *HMAP* would result in a new, higher Appropriate Management Level as well as reforms on the PMWHR and that a stay would only maintain the status quo. The *HMAP* in effect at the time of the appeal still authorized removal based on federal law and, in fact, would arguably result in more horses being subject to removal. The administrative judge denied the petition for a stay.[5] The 2009 *HMAP* remained in limbo, however, because an appeal was also pending. A second administrative judge affirmed the BLM decision on January 15, 2010, allowing the *HMAP* to take effect.[6] "Central to this appeal are various professional judgments regarding technical issues of genetic viability. The well-established principle of deference to expert opinions is, therefore, of particular relevance here. We have long held that BLM is entitled to rely on the professional opinion of its experts, where it concerns matters within the realm of their expertise and is both reasonable and supported by record evidence."[7]

The administrative judge went on to observe that the BLM had considered how to maintain a high level of genetic variation through other means, especially by preventing loss of bloodlines, retaining and increasing the core breeding population of five- to ten-year-old animals, and managing for an even sex ratio. Both Interior Board of Land Appeals orders came after the fall 2009 removal. Several outside wild horse activist groups were on hand to protest the roundup, but the only concrete effect of the appeal was to delay implementation of the range-improvement projects in the *HMAP*.

The 2009 *HMAP* reflected a new paradigm for on-the-range management of wild horses. Previous plans (1984 and 1992) had focused on forage-utilization studies and a periodic census of population numbers to calculate the Appropriate Management Level; however, the new *HMAP* placed greater emphasis on implementing steps to achieve the BLM "Standards for Rangeland Health." The idea behind the new plan was to adjust numbers, possibly upward, based on an evaluation of range-improvement projects. Amendments in 1976 and 1978 to the 1971 Wild Free Roaming Horses and Burros Act had placed greater emphasis on preventing deterioration of rangeland health from overgrazing using periodic removals. Ideally, proactive steps, combined with a scientifically managed fertility-control program, would alleviate grazing pressures and reduce the number and scale of wild horse removals.

The Center Joins a National Debate About Wild Horse Policy

The Billings Field Office published a gather-and-removal plan for the fall of 2009 and, despite the appeal of the *HMAP*, proceeded under existing authority provided by the 1984 *HMAP*, as amended in 1992. The Environmental Assessment stated, "The Proposed Action is needed to restore the wild horse herd numbers to levels more consistent with a thriving ecological balance and to remove or relocate wild horses from outside the PMWHR."[8] The EA noted that the area had experienced years of drought, and excess wild horses were allowed to remain on the range, exacerbating the effects of the drought. The last removal was held in 2006, and the number of adult wild horses on the range was estimated to be almost two hundred. The proposal was to gather most of the herd and selectively remove up to seventy adult horses.[9] The BLM proposed to conduct the operation after the foaling period in the fall.

The BLM based its decisions on whether to remove or release individual animals, gathered by helicopter drive, based on several criteria, but genetic representation (kinship) was a strong factor, in addition to age and genetic defects, such as hernias. The center had submitted a list to the Billings Field Office during the public comment period for the EA/Gather Plan, with genealogical data for each young horse born from 2004 to 2008. The center used the following criteria to recommend keeping specific animals:

Offspring of stallion either deceased, unlikely to reproduce or from rare bloodline
Offspring of mare either deceased, unlikely to reproduce or from rare bloodline
Offspring of rare stallion-mare combination
Mare reproducing with stallion from rare bloodline[10]

The center's recommendations favored removing young horses, born from 2004 to 2008, because they were easier to adopt. If there were siblings from the list still on the range, the center called for removing the younger horse in most cases. The list also noted horses with excellent Spanish type, characteristic of the Pryor Mountain Mustang, as a basis for keeping particular horses. The center's assessment of phenotype was similar to an evaluation provided by Dr. Sponenberg prior to the gather, and there were only a few occasions when the center's recommendations based on kinship conflicted with Dr. Sponenberg's report to the BLM.[11] That report concluded that previous management decisions by the BLM, favoring retention of the most typical horses, had solidified the Spanish characteristics of the herd.[12] It was issued before genealogical data would reveal evidence of crossing with outside introduced horses.

Ensuring that historic bloodlines were represented on the range was a new approach to managing the characteristics of the Pryor wild horses. The kinship method helped to preserve the Colonial Spanish horse type already established in the herd, as well as promote genetic diversity. The horses removed from the range were close to the center's list, but there were some substitutions when the contractor could not catch all the horses on the roster. Contrary to recommendations by the center and Dr. Sponenberg, the BLM removed all of the family bands roaming in areas of Custer National Forest outside of the PMWHR boundary, because they were in trespass and likely to return to the same areas even if relocated to the range. A few bachelor stallions representing all of the bloodlines on the Forest Service side did manage to escape. The final number removed during 2009 was twenty-eight horses from the PMWHR and twenty-nine horses from the Forest Service.

The participation by the center in the 2009 gather and removal followed in the tradition of the Pryor Mountain Mustang Association. Lovell advocates had usually understood the need to reduce the size of the herd population, even if they regretted the need to do so. A history of identifying specific horses to keep or remove, as well as assisting with adoptions, provided the context for the center's involvement in 2009. As noted in the center's winter 2009 newsletter: "The Pryor Mountain Wild Mustang Center volunteered to assist the BLM with the gather by having Director Matthew Dillon available for horse observation, identification and documentation. Dillon ensured that documentation and sorting were accurate and that females received the proper fertility control treatment."[13]

Although most wild horse roundups arouse public controversy, this particular one was the largest in the history of the PMWHR. Virtually all horses in the wild herd were gathered, or driven, by helicopter to holding corrals at the base of the mountain, and from there nearly sixty horses were removed and put up for adoption. The center's director had a key role because of his scientific education and ability to develop kinship charts for more than 90 percent of the Pryor herd. His weekly trips to the range and knowledge of the bands kept families together and helped to reduce stress on the horses. He also maintained a daily blog, including photographs, updating concerned friends of the center about the status of the daily gathers from September 3 to September 9 and posting photographs and names of the horses placed for adoption.

A series of educational blogs starting shortly after the 2009 gather and removal were based on historical and scientific documents and distilled otherwise complex information about the environment of the Pryor wild herd; the calculation of Appropriate Management Levels, or population limits; and the reproductive physiology of temporary fertility control, or PZP, into digestible educational modules. These blogs included maps, diagrams, and citations to research and countered criticism by outside wild horse activists of the PMWHR removals and fertility control administered to mares during the gather. The Cloud Foundation in particular was opposed at that time to PZP because of concerns that mares would experience continuous estrous cycles, leading stallions to become more competitive beyond the breeding season. TCF also favored a hands-off approach to population control and argued that the BLM should allow the herd to self-regulate its size.

Sharp differences between the center and TCF surfaced during and immediately after the 2009 gather and removal. Previously, the board had declined to be coplaintiffs in the lawsuit filed by TCF to expand the Forest Service portion of the range. Lovell advocates had always chosen to work with, not against, the BLM, but the new director's education added a different dimension to this consultative approach. BLM policy decisions were informed by science, and Matt Dillon spoke that language. His recommendations about which horses to remove during the 2009 gather and removal required an understanding of animal genetics and reproductive behavior. The list was specific and based on detailed analysis of wild horse pedigrees. As such it was an informed response, within the scope of the proposal, and was therefore data that the Billings Field Office could use in implementing its gather plan.

The gather and removal in 2009 was a microcosm of what was happening across the western states. The BLM planned to bring wild horse herd populations across the West to their Appropriate Management Levels by conducting multiple large-scale roundups. Holding facilities for wild horses were bursting at the seams, consuming a higher portion of the BLM operating budget; however, there were no plans to introduce fertility control to on-the-range management. In fact, the BLM was thinking about creating wild horse preserves to hold even more wild horses. This approach was unsustainable, because the wild horse populations on the range would just continue to increase without fertility control.

The center joined with another local wild horse advocacy organization, Friends of a Legacy, or FOAL, from the McCullough Peaks Herd Management Area near Cody, Wyoming, to send a letter, urging the BLM to declare a halt to nonemergency gathers until the agency could develop a new plan for managing wild horse herds:

> A priority for wild horse management would be for herds to be totally managed through natural means, such as predation. However, due to the effect that development in the West has had on the proper functioning of ecosystems, there are many areas in which artificial management must occur. The Secretary proposes the use of fertility control to reduce the growth rates of wild horse populations. We feel that a properly designed fertility control program should play a significant role in the management of wild horse populations; it is time for wild horse management strategies to be modernized.[14]

In addition to the BLM's strategic plan to improve management of the program through administrative changes, a major bill had passed the US House of Representatives and was now in the Senate: the Restore Our American Mustangs (ROAM) Act. The Senate bill was sponsored by Senator Mary Landrieu from Louisiana and sought to reform key structural features in the original law, such as the existing requirement that herd management areas be confined to locations in which the BLM found wild horses at the time of the act's passage.

That requirement limited the BLM's options as numbers increased in existing management areas. In the absence of systematic fertility control, the only other management tool available was gather and removal. ROAM aimed to change that requirement by proposing to require the secretary of the interior to "identify new, appropriate rangeland for wild free-roaming horses and burros, including use of land acquisitions, exchanges, conservation easements, voluntary grazing buyouts and agreements with private landowners to allow for federally supervised protection of wild horses and burros on private lands, except that the Secretary shall assess the effects of new range for wild free-roaming horses on rangeland health."[15]

The US Senate never took up ROAM, in part because Senator Landrieu had to address the crisis in New Orleans caused by Hurricane Katrina; however, efforts by the BLM to seize the initiative and propose to Congress a new strategic plan allowing it to reform the program using internal, administrative changes rather than a politically controversial legislative overhaul were a more likely reason ROAM stalled in the Senate.

PMWMC Volunteers Help to Implement the New *HMAP*

Once the appeal filed against the 2009 *HMAP* by the Cloud Foundation was settled in favor of the BLM, the PMWMC began helping with range-improvement projects. During the summer of 2010 board secretary Kathy Johnsey and her husband, Bill, worked with volunteers to install new water sources, or "guzzlers." As reported in the newsletter, new water sources could help to stabilize conditions, especially at the top of the mountain: "A goal of this plan is to encourage the wild horses to spread out across the Range more. This could allow for a higher population of wild horses to live on the Range. Working alongside members of the Montana Conservation Corps and the Bureau of Land Management, PMWMC volunteers provided a total of nearly 100 hours of labor on this first guzzler."[16]

Another step in implementing the new *HMAP* was helping Billings Field Office personnel with the fertility-control plan. In December 2010 the Billings Field Office published the "Pryor Mountain Wild Horse Range Fertility Control Environmental Assessment."[17] The plan called for a five-year program to supplement previous vaccinations. "The Proposed Action is a fertility treatment program in order to implement the 2009 PMWHR *HMAP*. The program would start in 2011 and last through 2015. The fertility-control program would consist of the administration of native PZP applied through remote darting in the one year liquid dose. The program would be designed to treat mares 2, 3, 4 and ages 11 through 20+. Mares ages 5–10 would not be treated."[18]

The plan estimated that eleven mares out of the eighty-two in the population might be permanently sterile, because they would have been treated with PZP-22 pellets during the 2009 gather and removal, plus up to five annual doses of the native or liquid PZP. The plan noted, however, that these mares would be unlikely to produce foals because of their advanced age.[19] The objective of the plan was to allow young mares to mature before becoming pregnant, but then consider them as part of the core breeding population. Because mares were vaccinated based on their age cohort rather than their bloodlines, the importance of considering representation on the range during any future removals would become paramount. Both PZP and removals were "tiered" to the *HMAP*, which set the Appropriate Management Level contingent on fertility control and also established criteria for future removals, including the representation of bloodlines on the range.

Remote darting involved finding individual mares, approaching them slowly, shooting the dart accurately from thirteen to fifteen yards away, and then retrieving the dart after it hit the mare in the hip and fell to the ground.[20] This task went to the center's director, Matt Dillon, because he knew each horse on the range and was trained in the application of native PZP. His role during 2010 was to coordinate volunteer efforts and to assist with implementation of the new *HMAP* as a living document. The action steps were organically related: the health of the wild horse herd depended on range improvements; conservation of forage and water, in turn, depended on maintaining a sustainable number of horses on the range; and controlling the numbers of wild horses depended on coordination of temporary fertility control with periodic, small-scale removals based on demographic and genealogical data in order to preserve distinct Pryor bloodlines.

In 2010 the Billings Field Office recommended Matt Dillon for a national BLM award based on his leadership in helping with the implementation of the new *HMAP*. The center noted in its newsletter that his leadership reflected the values of the board: "This spring, the Pryor Mountain Wild Mustang Center Director, Matthew Dillon, was honored to receive the Bureau of Land Management's 'Making a Difference Award.' Dillon received this award for the work he has done in organizing and providing volunteer assistance on range improvement projects as well as wild horse data collection. These actions reflect the philosophy of the PMWMC, which seeks to work with, and not against, the BLM in finding ways in which the Pryor Mountain Wild Horse Range, and the wild horses living there, can be properly managed."[21]

Many wild horse activists were opposed to any form of management for wild horse populations. The Cloud Foundation had appealed the *HMAP*, and now objected to the fertility-control plan on the grounds that continued use of PZP would cause permanent sterility in mares. Other concerns were that PZP would have a negative effect on social interactions, because immunized mares would continue having heat cycles throughout the summer and fall. Continuous estrous cycles would, in their view, lead to more aggressive behavior between stallions competing for mares and even encourage mares to be unfaithful to their own band stallions. The fact that Matt Dillon had been so active assisting the BLM during the 2009 gather, 2010 range-improvement projects, and remote darting of mares on the range using native PZP made him a target of fierce criticism from many wild horse activists.

The Cloud Foundation was especially concerned about the PMWHR because it was the site of the PBS *Nature* documentaries filmed by the organization's volunteer executive director, Ginger Kathrens. The series followed the life of a palomino roan stallion, Cloud, from birth to maturity as the head of a large family band on the PMWHR. The documentaries created a large and loyal following that read stories on TCF's website that were critical of BLM policies and sent comments to the BLM, objecting to both removals and fertility control. Visitors to the center sometimes asked specifically where they could see "Cloud's herd," and they seemed less interested in the Colonial Spanish heritage of the wild horses.[22] Relations between the center and the Cloud Foundation became strained during this time because of the opposing beliefs of the two organizations, one with a local mission and the other with a national mission, and their governing boards.

The main source of tension between the center and outside wild horse activists was its consultative relationship with the BLM. Historically, Lovell advocates set the tone by keeping the lines of communication open during the tense period leading up to the creation of the PMWHR and avoiding outside organizations that were suing the BLM over various allegations. Their philosophy of working with, not against, the federal government led to indictments by outside observers: Lovell advocates were accused of exploiting the Pryor wild herd mainly to bring more tourist dollars to the town, folks breeding Pryor Mountain Mustangs were supposedly joined in a profit-making venture, and leaders of the earliest Lovell advocates allegedly looked the other way while the BLM committed atrocities during the first roundups. Center board members tolerated these attacks but bore "scars" for years.

The board's president, John Nickle, had been involved with the Pryor wild horses since first coming to Lovell in the late 1960s as town manager. The mayor at that time, Cal Taggart, later served on the center board. Several center board members were the initial directors of the Pryor Mountain Mustang Association when it was reorganized in 1992. They included John Nickle and the Reverend Floyd Schwieger; Jim Edwards, who owned the Sweet Medicine Wild Horse Sanctuary in Fishtail, Montana; and Earl Lukkes, who owned several Pryor Mountain Mustangs registered with the breeders association. Although Jim Edwards had adopted ten Pryor mustangs from the BLM in the early 1990s, he did not register them with the breeders association. Thus, when the center was in the planning and construction phases, the makeup of the board included several of the earliest Lovell advocates.

Those early Lovell advocates worked closely with Dr. Phil Sponenberg and the Billings Field Office during the early and mid-1990s to preserve the Pryor Mountain Mustang by scoring individual horses on the range and recommending which ones to keep in the wild herd because of their excellent Spanish phenotype. Later, during the planning phases for the center, board member Nancy Cerroni took photographs of each wild horse and brought them back to Floyd Schwieger, who taught her about kinship patterns. Her son, Matt Dillon, continued to develop the center as a locus of expertise about the Pryor wild horses by developing and maintaining the genealogy project. Thus, the center's consistently moderate position on policies affecting the PMWHR was based on not only the "Lovell way" of getting along with neighbors but also the firm conviction that, as stewards of the range and its unique wild herd, they needed to base their positions and actions on the best available science and avoid projecting human emotions onto wild animals.

The center developed a positive relationship with wild horse specialist Jared Bybee because of his commitment to bringing twenty-first-century science and management to the PMWHR. This positive consultation served only to further aggravate outside activists who distrusted the BLM. In January 2011 Matt Dillon resigned as director of the PMWMC to assume a position as an environmental scientist for a local mineral company. He continued to work with Billings Field Office personnel as a board member to monitor the wild herd, administer PZP, and maintain the genealogical database. Nevertheless, his resignation was a major loss to the center. His scientific background in technical areas, such as fertility control and genealogical data, would be virtually impossible to replace. A small nonprofit organization, operating on contributions from visitors and other supporters and governed by a board of local volunteers, was vulnerable to even the slightest changes, much less the resignation of its founding director.

The Future of the Pryor Mountain Mustang Appears Uncertain

By the summer of 2011 the board had appointed Lori Graham to the position. She had been a board member and was familiar with the history of the PMWHR, but her background and interests reflected a new generation of volunteers. She worked as a wildlife rehabilitator in Florida before moving to Lovell with her husband in 2003. Her love of horses began at an early age, and she soon acquired two Pryor Mountain Mustangs, one raised by Dale Hartman and one originally adopted from the PMWHR and rescued from the adopter. Another new member to the board, Diane Granger, was an animal defender in New Jersey before she moved to Lovell. She and her (now deceased) husband adopted two Pryor wild horses in 2001. These new board members formed a deep emotional bond with the Pryor wild horses, but because they were new to Lovell and unfamiliar with the long institutional history of the PMWMC and BLM, a new perspective on the mission of the center was, perhaps, inevitable.

The center had held its first "Meet the Mustangs" event in 2004, before groundbreaking on the building the following year. That event continued every summer until 2010, giving the public a chance to see and touch the Pryor Mountain Mustangs bred and owned by Dale and Daphne Hartman. Although the Hartman family no longer pastured their horses on the center's property, they trailered them to Constitution Park in Lovell to show off their unique Colonial Spanish–type conformation, color, and markings. By 2011, however, they were planning their retirement, and the center held an annual open house after that but without a Meet the Mustangs event.

In the early 1990s when the PMMBA formed as an organization separate from the Pryor Mountain Mustang Association, the missions and memberships of the two entities overlapped to a large extent. Board members from both organizations shared a passion for preserving the bloodlines of the Pryor Mountain Mustang as a strain of Colonial Spanish horse by selectively retaining horses both on the range and on breeding farms with high Spanish phenotype scores, including conformation, color, and markings. The *Official Horse Breeds Standards Guide* states: "The PMMBA recognizes the unique genetics of this important breed and attempts to preserve it by selectively breeding adopted Pryor Mountain Mustangs. The number of PMMBA horses has been recognized by the American Livestock Breeds Conservancy as critical, with fewer than two hundred registrations each year and an estimated global population of less than two thousand. The Pryor horses are on the critical list along with nine other Colonial Spanish Strains."[23]

The Pryor Mountain Mustang was a unique part of the area's western heritage, and it captured the imagination of the earliest Lovell advocates, providing them with a fascinating narrative that attracted visitors to their community and added weight to the argument for setting aside the first public wild horse refuge. The original goal of the PMMBA was to create a gene pool containing rare Pryor bloodlines in case of a catastrophic event on the PMWHR. During the 1990s members of the Pryor Mountain Mustang Association advocated for preserving the wild herd, but also adopted or purchased Pryor Mountain Mustangs and shared with their friends in the PMMBA an enthusiasm for the versatile performance of these remarkable animals as trail horses, stock horses on cattle ranches, and performers in the show ring. For members of the two organizations, the line between the Pryor Mountain Mustang as wild horse and domestic livestock animal was blurred. The common goal was preservation of rare Pryor bloodlines. By 2009, however, leaders of this common cause had died or retired or were planning their retirements.

Many Lovell residents, including new center board members, adopted Pryor wild horses during BLM removals, but did not necessarily register their horses with the PMMBA. By 2009 interest in preserving a domestic gene pool of Pryor bloodlines through breeding seemed to have waned, in large part because the horse market was suffering from the recession. The center's director, Matt Dillon, had wanted to work with the PMMBA to update the Pryor family trees.[24] There were likely to be fewer foals from mares owned by breeders, however, because of retirements and cutbacks due to the weak horse market. The common vision held by the Pryor Mountain Mustang Association and the breeders association to preserve Pryor Mountain Mustangs was significantly affected by these factors.

As early as 2008 one breeder foresaw the fate of the common vision held by so many people who loved the Pryor horses for their unique heritage and versatility in work and sport activities. Virtually everyone who has bought or adopted a Pryor horse speaks about their uncanny ability to bond with people. This breeder expressed his concern about the implications of using fertility control on the Pryor wild herd, including the possibility of fewer removals and opportunities for adoptions: "Add this to the round up and adoption no longer being an option, the Pryor Mountain Mustang will no longer be available to the public and the only option of keeping this beautiful rare breed going will be through private breeders, of which there are only a hand full. Our goal as a private breeder is to preserve and protect these horses and the genetic balance that now exists."[25]

Of course, the largest gather and removal in the history of the PMWHR happened just a year later, but the long-term objectives of the 2009 *HMAP* included fertility control and the need for fewer and smaller removals. The "kinship method" favored by center director Matt Dillon thus took on greater importance as population growth stabilized, because preserving rare bloodlines now depended less on retaining individual horses with high phenotype scores and more on preserving representation of all existing bloodlines in the wild herd. Complicating matters, however, was new evidence from the center's genealogy project that some offspring of the Wyoming and Burns, Oregon, horses introduced in the 1980s had survived on the PMWHR despite prompt removal of the parents and that there had continued to be crossing with rare Pryor bloodlines. It would be difficult now to separate those bloodlines.[26]

This development led Dr. Phil Sponenberg to state flatly that he could no longer give blanket validation to the entire wild herd as being Spanish. In his view a unique genetic resource was being lost.[27] His work with the Pryor Mountain Mustang Association and the breeders association during the 1990s to identify and recommend keeping Pryor wild horses on the range based on Spanish phenotype scores had taken place before the availability of genetic data from the genealogy project. In hindsight the effects of crossing between bloodlines may have accounted for a number of horses on the PMWHR being scored as "off-type."

Of course, the wild horses saved by the earliest Lovell advocates were far from genetically pure, suggesting that the Spanish type had survived crossings with homesteader horses. Similarly, crossings with outside introduced horses and rare Pryor bloodlines produced many animals with the conformation, colors, and markings associated with the Pryor Mountain Mustang: small stature, tapered rather than bulky musculature, dun or mousy-gray color, and primitive markings, such as dorsal (back) stripes, zebra stripes on the legs, and black stripes or bars across the withers and shoulders. Basing removals on maintaining representation of all the bloodlines on the range could not guarantee the survival of Colonial Spanish bloodlines, but the database could help identify stallion-mare combinations that should be preserved.

The BLM Conducts a Bait-Trap Removal

The BLM fertility-control plan was less aggressive than the program on Assateague Island where each mare was allowed to produce only one offspring. Such an approach would have been difficult to manage on the PMWHR. The difference in scale of the two ranges was reason enough, but concerns about preserving rare Pryor bloodlines meant that neither the BLM nor the center wanted to risk the consequences of a catastrophic event. As a result the numbers of Pryor wild horses began to grow following the 2009 gather and removal, despite initial efforts to limit population increases through implementation of the fertility-control plan. While the need for another gather, even on a small scale, was inevitable, BLM wild horse specialist Jared Bybee was determined to avoid a helicopter drive.

> The purpose of the Proposed Action is to meet the goals and objectives of the PMWHR *HMAP* by maintaining the AML. The action is needed because BLM has determined excess wild horses are present on the range. The current population of wild horses is approximately 170 animals. A helicopter survey conducted on March 12, 2012 by the BLM and NPS personnel classified 166 wild horses present on the range. The BLM has measured heavy and severe utilization of vegetation forage species from wild horses grazing

on the upper elevations and lower elevations of the PMWHR. The *HMAP* identified the AML at 90–120 wild horses as the carrying capacity in order to maintain ecological stability on the range, and protect the range from deterioration.[28]

The plan called for the removal of thirty to forty excess animals, targeting the one- to three-year-old cohorts using a combination of bait and water trapping.[29] According to the Environmental Assessment, the proportion of the herd falling into the one-, two-, and three-year-old categories constituted 40 percent of the current population. *HMAP* removal considerations targeted horses with low Colonial Spanish–type scores, as well as horses under five years old that were genetically well represented on the range.[30] The plan therefore classified one-, two-, and three-year-old horses into tiers based on how well their bloodlines were genetically represented on the range, as reported by Matt Dillon.[31] The plan managed for kinship, sex ratio, and phenotype, although it identified Spanish type as a secondary consideration: "Since young animals are more difficult to properly score phenotypically and the majority of Type 1 and Type 2 Spanish Phenotype (as defined in the *HMAP*) animals (over 3) are still on the range post 2009 gather, the Spanish characteristics are already retained in the herd."[32]

The plan for bait trapping was to lure family bands into portable steel-paneled corrals. These traps were set in several areas at upper and lower elevations of the range frequented by the horses. First, the "bait" (food or salt, or both) was set out in the designated spot; then, after the horses began coming to those spots, panels were set up on two of four sides in order to allow horses to pass through freely until they became accustomed to the traps. Once the trapping started, a BLM or NPS field staff member had to stay on site to trigger the automatic gate when one or more family bands wandered in and then to separate the young horses targeted for removal before releasing the rest of the horses.[33]

The removal took the entire summer to complete.[34] In the end the BLM removed thirty-eight adult horses and six foals for adoption at the Britton Springs corrals. Field staff, headed up by wild horse specialist Jared Bybee, had to wait for the right family bands and separate the correct horses. That step alone took several weeks. Small groups of young horses were then trailered to the bottom using Crooked Creek Road, a three-hour trip skirting the western boundary of the range. Multiple trips added to the length of the removal. By September 8 all of the horses (including mare-foal pairs) were ready for adoption, and all went to adoptive homes. As described in a local newspaper account:

> The traps were first set up on the lower elevations of the range on June 20. Those horses proved the most difficult to trap since they were more skittish about the sight of humans. On July 9 the traps atop the Pryors were set up. Those horses show little fear of humans and have been much easier to trap.
>
> Unfortunately for BLM two bands—those led by studs Duke and Jackson—have largely avoided the trap sites. "We're down to two bands that we need to get horses out of, but they're the two biggest bands and the studs are reluctant to bring them in," Bybee said.
>
> On a recent afternoon, however, Duke finally showed up at the trap site set up atop the Pryor Mountains above Krueger Pond. Inside the corral, Duke whinnied to his mares hiding in the spruce trees hundreds of yards away as a band of young studs hovered close to the trap site, eagerly waiting for a food handout.[35]

The removal aroused much less controversy compared to 2009. The scale was much smaller, because the size of the herd had already been substantially reduced. Unlike several years ago, the process was more transparent to the public. The BLM Billings Field Office posted maps of the trap sites and opened those areas to the public, subject to safety restrictions. It was possible for wild horse activists to photograph, videotape, and blog at each step of the way. The corrals at Britton Springs were also open to the public so that the condition of the horses was visible to observers. In 2009, by way of contrast, the range at the top of the mountain was closed to the public, and there was a small observation site at a distance from the areas where the helicopter drove horses through the chutes into holding pens.

The 2012 removal was "low impact" for both the public and the horses. According to the BLM website, a total of 146 horses were gathered in the traps, while the 38 young horses targeted for removal were taken to the Britton Springs corrals. Six young mares had foals with them, and a mare foaled in the corrals shortly after she arrived.[36] For the first time staff from the National Park Service partnered with the BLM. In his blog Matt Dillon wrote, "I believe that this gather had gone very well. The BLM has put a significant amount of time into this gather; about a month will have gone into it by the time it is over with. In this month's time, there had also been a pretty good amount of overtime. All this shows, though. There haven't been any injuries during operations. The horses have been able to be strategically removed so that bloodlines will not be eliminated. I believe that this could serve as a model for proper management."[37]

The national office of the BLM presented the Pryor Mountain Horse Gather and Adoption Team, headed by Jared Bybee, the Director's Team Accomplishment Award. The award reflected the intensive effort required to manage a wild horse bait-trap removal, as well as maintain the public's trust. While some wild horse activists will always find it hard to trust the federal government, especially the BLM, many others commended Jared Bybee and his team for the care they took with the horses.[38] Their positive comments showed that the downward public relations spiral experienced by the BLM in other states could be reversed by combining sound wild horse management with improved transparency to the public. The BLM's press release stated, "This Award recognizes this year's tremendously successful Pryor Mountain wild horse gather and adoption. Thanks to outstanding public engagement, sound decision making, and cohesive team work, this gather and adoption were completed safely and with positive public support."[39]

The role played by the center was a major factor in the success of the 2012 bait-trap removal, not only because of its small scale and low impact but also because the genealogy project allowed the BLM to identify which young horses should be kept on the range. Although no one wanted to see more horses removed, especially the center, a small-scale gather meant that the BLM could avoid a larger helicopter drive in the future. Thus, the team that won the award could have included the center, in recognition of years spent volunteering to assist the Billings Field Office with monitoring of the herd and range-improvement projects, as well as stepping up to educate the public about the complexities underlying the management of this unique wild herd.

Chapter Five

Looking to the Future

The individuals and motivations for becoming involved with the Pryor wild horses have changed over the years and will undoubtedly continue to evolve in the years to come. What has remained constant throughout this history, however, is a commitment to the preservation of the Pryor wild herd. No other community-based nonprofit organization can claim sustained forty-five years of grassroots advocacy on behalf of a wild horse herd in its immediate vicinity. The challenges that the center faces in the near future are those confronting many community-based nonprofit entities: complex issues, combined with scarce financial and human resources, will encourage them to form networks with other nonprofits. The center maintained its autonomy, establishing a unique identity and relationship with the BLM. Its board and next director will therefore have to walk a fine line between maintaining the center's mission and establishing new networks.

The Future of Pryor Mountain Mustang Breeders

After Dale and Daphne Hartman made plans to retire and travel, the center assumed responsibility for the Pryor Mountain Mustang Breeders Association registry records and filed papers forming the Pryor Mountain Mustang Registry Association on June 12, 2011. A longtime member of the original PMMBA, Bess Carnahan became the new registrar. In August 2012 she registered Pryor Mountain Mustang no. 219. The list includes all horses ever registered, including those horses that have died, so the number has remained basically the same since 2007 when the PMMBA newsletter reported 194 Pryor Mountain Mustangs.

The original reason for forming the PMMBA was to preserve the Pryor Mountain Mustang as a unique strain of Colonial Spanish horse by conserving bloodlines in case of a catastrophic natural event on the range. The association promoted interest in the Pryor wild herd by participating with the center in its annual Meet the Mustangs event and with the BLM at its Britton Springs adoptions. Its members also joined efforts during the 1990s to advise the BLM during gathers and removals, emphasizing Spanish phenotype scores developed by Dr. Phil Sponenberg. After 2009 the Billings Field Office began to emphasize the kinship method, as well as Spanish phenotype, but until director Matt Dillon compiled data on the bloodlines of the horses, phenotype score was the most objective method available for making decisions about removals during periodic gathers.

Although there have been a number of PMMBA members since the early 1990s when the association was formed, only a handful of breeders remain active. Besides Bess Carnahan, who lives in Nebraska, there is a US breeder at Arrowhead Ranch in Massachusetts who began in the 1990s with two mares sired by Dunbar and purchased from Dale Hartman. The owner, Red Tolman, bought his stud from Cathy Nunn, Dale Hartman's daughter, and a third mare from the Reverend Floyd Schwieger. By 2012 there were seventeen registered Pryor Mountain Mustangs roaming in large pastures bounded by woodlands on the outskirts of Baldwinville in western Massachusetts. He recently gelded his stallion, because he is approaching retirement age, has no family members to take over the ranch, and has been unsuccessful in breaking even from sales of his horses. His hope was to spark interest in the Pryor Mountain Mustang among New Englanders, but the economic crisis and downturn in the horse market killed that dream. He is now trying to find good homes for each horse.[1]

The breakup of Arrowhead Ranch would be significant for the future of breeders, because the horses there are the last remaining domestic herd from the foundation stock acquired by Dale Hartman from the Tillett ranch. Although no Pryor wild horse that was removed from the range ever ended up in a BLM long-term holding pasture, adopted animals are rarely registered now.[2] There is no longer a separate membership association based in Lovell, and, compared to the 1990s when there was a strong tradition of participation by members in the Mustang Days parades, Meet the Mustangs, and mustang breed and performance classes at local and regional horse shows, this part of the forty-five-year history may be coming to an end. The Colonial Spanish heritage of the Pryor horses was a major reason there was so much public attention to, and support for, the creation of the PMWHR. The futures of the PMMBA, the center, and the Pryor wild herd are therefore connected in many ways.

For the first time the center adopted two young Pryor wild horses after the 2012 bait-trap removal. Visitors to the center during the summer of 2013 could once again observe the unique coloring, markings, and gentle disposition of Pryor Mountain Mustangs. The Hartman horses that once grazed in the large pasture behind the center were registered, "gentled," and accustomed to humans. The board is now discussing a plan to move the newly adopted, young Pryor horses to the property of one of its members who already has the facilities and time to work with them. The large pasture behind the center proved to be unsuitable for the newly adopted young wild horses, and the small enclosure and shed were designed only as temporary shelter. The board is discussing the possibility of rescuing Pryor horses whose owners are no longer able to keep them. These older Pryor Mountain Mustangs are socialized to being around people and living in a pastured environment.

Ironically, a new generation of Pryor Mountain Mustang breeders has started in France, close to the Iberian Peninsula, where the ancestors of the Pryor mustangs originated. They own registered Pryor horses and want to protect this unique strain of Colonial Spanish horse. Like the earlier Lovell breeders, their goal is to preserve the Pryor Mountain Mustang and to educate as many people as possible, so that there will be public support for the wild herd. As breeder Cedric Berthouzoz explained, "Here in Europe, mustangs are 'exotic.' They are not seen the same way like in the U.S. where many people see them as vermin."[3] His interest in adopting a Pryor wild horse began in 2005 when he discovered the "Cloud" documentaries. He had hoped to adopt a wild horse, but BLM rules require that adopters be US citizens. So he found Bess Carnahan by searching for Pryor Mountain Mustang breeders on the Internet and asked her many questions before purchasing a registered stallion from her in 2008 and also a mare in 2012. His mare is now bred to his stallion.[4]

In contrast with breeders in the Lovell area who emphasized the link between the Pryor Mountain Mustang and the heritage of the West, Cedric's interest is in Spanish history and culture. Returning the Pryor Mountain Mustang to its European home where its ancestors were bred as battle mounts for the conquistadores is one major reason he is interested in these horses. Although he admitted that he would not be able to make a living as a breeder, his interest motivated him to make a considerable personal investment in bringing his stallion and mare to France. He has studied the history of the Iberian horse from its Ice Age ancestors in Europe and forward to present-day Haiti, where a Swiss journalist covering the earthquake photographed a small horse bearing a striking resemblance to a Pryor Mountain Mustang. In fact, the Caribbean islands were home to the conquistadores before their exploration and conquest of Mexico.[5]

A second breeder, Florence Mouninou, moved from France to Livingston, Montana, in 1992. She followed the Pryor wild horses, learned about their history from the Reverend Floyd Schwieger, and adopted several horses through the BLM. In 2011 she brought her Pryor Mountain Mustangs home to France, including a stallion, all of which are registered. She is now building a ranch in France where she plans to start a breeding operation for reasons similar to her American, especially Lovell, counterparts: "It's important to preserve this unique breed and heritage for future generations."[6] Like the earlier breeders, Florence Mouninou wants to preserve the Pryor Mountain Mustang by establishing their authenticity through the registration process and maintaining a gene pool in the event the wild herd would be extinguished by a catastrophic event.[7] These two French breeders are carrying on the mission of the original Pryor Mountain Mustang Breeders Association, while also closing the historical loop by bringing horses back to the Iberian Peninsula.

The Pryor horses now living in France were either adopted from the BLM or sold by Bess Carnahan, so their genealogy is likely to include rare Pryor bloodlines. Another small herd living off the range came originally from bands roaming in Custer National Forest in areas deemed by that federal agency to be outside of the PMWHR boundary. In the 2009 removal virtually all of the Forest Service family bands were taken from that part of the range. The Cloud Foundation then acquired all four family bands and cares for them on land leased from a private ranch north of Livingston, Montana. TCF acquired these bands because it wanted a genetic reservoir in case of a catastrophic event on the range. It is a reproducing herd, although mares that have had foals are on native (one-year) PZP.[8] TCF has a lawsuit against the Forest Service, claiming that the areas formerly occupied by these family bands should be included as part of the range.

Another potential new breeder is Steve Smith, from Henry, Nebraska, who is certified in equine-assisted learning and works as equine therapy supervisor at the Scotts Bluff County Detention Center in Nebraska. He is training several Pryor wild horses adopted by Bess Carnahan, and he has found that even Pryor horses recently removed from the range have an uncanny ability to assist him in his work with private clients at his ranch. Pryor wild horses are unafraid of people, but are also more sensitive, and respond more quickly than domesticated horses, to the body language of humans. He has been able to use them in assisting troubled youths in learning about respecting boundaries by asking why a horse approaches—or flees—from a particular gesture or body movement. His clients become aware of how their behavior can affect human and nonhuman beings. His work with Bess Carnahan's horses has led him to consider becoming a Pryor Mountain Mustang breeder himself.[9]

The Pryor Mountain Mustang has been listed as a critical Colonial Spanish strain by the American Livestock Breeds Conservancy, defined as having fewer than 200 annual registrations in the United States and a global population of fewer than 2,000. It is also one of only four strains with a feral or wild origin.[10] There are presently only 220 *total* registrations and a historical annual average of 150 adults on the range. Even if all Pryor wild horses removed and adopted since the creation of the PMWHR were registered, the sum of that population would still be fewer than 2,000 animals. If, as Dr. Phil Sponenberg has stated, the Spanish heritage of the wild herd is diluted because of outside introductions and interbreeding, small herds in Nebraska, Massachusetts, Montana, and France may be among the only remaining bands descended from the original Pryor horses.[11] The future of the Pryor Mountain Mustang as a distinct breed may therefore rest with this new generation of breeders.

In fact, the history of the breeders association has always been controversial. Some wild horse activists have accused breeders of creating inhumane conditions by confining a number of horses in spaces that are too small and cramped for wild horses; selling Pryor horses without screening potential buyers, as would be required in a BLM adoption; and generally exploiting Pryor Mountain Mustangs as a profit-making venture. The demise of breeding operations in the United States through the retirement or death of the original PMMBA members is frankly greeted with relief by these critics. Importation of registered Pryor Mountain Mustangs to Europe has, to date, been greeted with greater interest and sympathy there than in the United States. That trend may change if veteran breeders, like Bess Carnahan, can recruit a new generation of US breeders, such as Steve Smith in Nebraska.

The Future of the Center's Consultation with the BLM

Center board members have a high degree of confidence in field office manager Jim Sparks and wild horse specialist Jared Bybee and acknowledge that the excellent condition of the wild herd as well as the increased amount and quality of forage are due to their twenty-first-century approach to on-the-range wild horse management. They have also come to trust these BLM officials because of incidents when individual horses have been reported injured or missing by the center. One story retold by board members is about Phoenix, palomino mare and mother of Cloud, with a sizable sore on her hip. After monitoring her for a short period, Jared Bybee gathered and transported her to a veterinarian for treatment. He then released her on the range but only after waiting for her band stallion, Teton, to appear. Incidents like this one show that Billings Field Office staff members are both caring and competent.

Lovell advocates have consistently chosen to work with, not against, the BLM. The 2008 Memorandum of Understanding reaffirmed that the BLM and the center "share many common goals for the management of wild horses."[12] It affirmed the long-standing relationship between Lovell advocates and the Billings Field Office, beginning with the Reverend Floyd Schwieger's detailed lists and photographs of the family bands and bachelors on the PMWHR and his monitoring of the herd throughout the year. After his death in 2005, this role was filled by the first director, Matt Dillon, who updated the horse list and developed the genealogy project. During the period when he was director, he also published frequent blogs designed to educate people about the scientific basis upon which the BLM based its decisions.

Monitoring the herd, in addition to public outreach and education, will require more scientific and technical expertise than Floyd Schwieger had when he kept his detailed records during the 1990s. The basis for management decisions on the range is also the target of more intensive scrutiny by wild horse activists who distrust the BLM and oppose both removals and the use of fertility control. The center tried to defuse this opposition through its public outreach and educational mission, but this effort backfired when outside activists accused Matt Dillon and the center of colluding with the BLM. Nevertheless, the center's consultative relationship with the BLM is based on its ability to maintain consistent and reliable databases about the herd. The frequent turnover in the director position since 2011 has been a major setback, but it is an opportunity to clarify the center's mission, vision, and values.[13]

The center's influence with the Billings Field Office stems from local expertise and the capacity to deliver demographic and genealogical data on which the BLM can base decisions. As a government agency authorized to manage wild horses in balance with the environment, BLM field offices must stay within their statutory parameters. When a field office consults with the public about gather-and-removal plans or fertility-control plans, its staff members rely on input from which they can base their decisions. Reading the public comments offered by many wild horse organizations and individuals makes it apparent that people have strong opinions about the BLM and will use the opportunity to comment on a range of issues of concern to them that are unrelated to the published proposal. The center's input is targeted to specific proposals and includes data that support a scientific rationale for a final decision by the BLM.

The history of Lovell's involvement with the Pryor wild herd and its consultation with the BLM shows just how important it is to have a center director or board member with the interest, time, and talent to develop detailed local knowledge that speaks directly to the specific issues at hand: How many wild horses are sustainable on this particular range? How should the field office decide how many and which horses to remove when the population exceeds the Appropriate Management Level? How aggressively should the field office use PZP? Answers to management questions such as these are defensible only when the underlying justification is sound. The BLM cannot delegate its legal authority to an outside organization; however, the comments of an informed community-based organization such as the center can influence the public management of a wild horse herd because of its local expertise. The center's niche has developed over many years of local expert consultation.

The Future of Support from the Lovell Community

The inspiration and support for the original Pryor Mountain Wild Horse Association came from the historical committee of the Lovell Chamber of Commerce. Its members saw the potential of the wild horse range, not only as a tourist attraction but also as a symbol of the community and its western heritage. Once these members formed the Pryor Mountain Wild Horse Association, the Lovell Chamber became involved with the campaign to save the wild herd. Hundreds of schoolchildren and adults sent letters to the Chamber in support of the campaign. After the Secretarial Order creating the PMWHR in 1968, public attention died down, both inside and outside Lovell. During the 1990s, however, when members of the Pryor Mountain Mustang Association were planning for the new center, the town as well as local banks and businesses loaned and donated funds and materials for the land and physical structure. Area residents also donated hours of in-kind services—a frontier-style barn-raising event.

Lovell, Inc., a nonprofit economic development corporation, recently held a workshop attended by seventeen people, mainly from Lovell, during which participants identified Lovell's "brand" or symbol distinguishing their community from other tourist, business, and residential destinations with natural beauty and recreational opportunities. Residents who attended the workshop agreed that Lovell "was not for everyone" who wanted to visit or move to the West. Compared to Cody, Wyoming, with its Buffalo Bill Center, shops, and "Wild West" reenactments, Lovell was a smaller-scale but less commercial version of the old West. The trademark, they agreed, should be authenticity, and promotional materials should read "Get Real," in the sense of getting real freedom, adventures, opportunities, experience, and beauty.

As the group discussed what image came immediately to mind when they thought of their community, they realized that it was the Pryor Mountain Mustang. The phrases associated with Pryor Mustangs were *wisdom, freedom, independence,* and *family bonds;* these phrases were also the same values held by the Lovell community. The workshop also talked about the unique Spanish lineage of the herd, the fact that people felt a bond with the wild horses because the center and BLM had given each Pryor horse a unique name, and the belief that humans could learn from horses. Although there may be only a core group of about fifty locals committed to the Pryor wild herd and to grassroots advocacy on its behalf, the potential exists to establish a new sense of connection between Lovell and the center.

According to Sue Taylor, Lovell, Inc.'s director, there are opportunities for networking with the center. The Mustang logo will be on all printed materials. Moreover, Lovell, Inc., can offer the kind of support once provided by the Chamber of Commerce. The board president is David Peck, publisher of the *Lovell Chronicle* and longtime supporter of the Pryor horses. The other board members represent important sectors of the Lovell economy, including the chief executive officer of the North Big Horn Hospital District, the representative from State Legislative District 26, and the town clerk-treasurer. The board is committed to small-business start-ups and drawing in new residents who share Lovell values; however, its members also want current residents to identify with the new brand. Their vision is community economic development—growth that is organic and blossoms from within

Lovell.[14] The Pryor wild horses are known and loved by people from around the world. The fact remains, however, that without a core group of local volunteers, it will be difficult to sustain the center, much less expand its current level of activities.

In the past local support for the center was built on personal relationships between board members and recruits. Just about every living past or present member of the board has a personal story about how the Reverend Floyd Schwieger, Dale Hartman, or John Nickle made a connection with them and saw how their talents and interest could contribute to the organization. Leadership by individuals, such as Floyd Schwieger, energized the original grassroots campaign to create the PMWHR, and leadership will continue to be essential for sustaining the center's mission, vision, and values. The center has been an effective local voice for the Pryor wild herd because it has avoided getting entangled in national wild horse politics. Its credibility with Lovell residents is, in large part, a function of how closely center values are aligned with the local culture. People are independent and suspicious of outsiders with a radical agenda. As one board member put it, they are not "tree huggers or horse huggers."

The center has remained autonomous from national networks of wild horse activist organizations. Its mission has been clearly focused on the PMWHR instead of spread out over other herd management areas in the West. Its archives, collection of books, permanent displays, and website are targeted to Lovell's involvement in creating the range, the characteristics of the Pryor Mountain Mustang, and the social dynamics of wild horses—all information that is unique to Lovell. The more this outreach targets Lovell residents, as well as visitors from other states and countries, the better the chance of recruiting more Lovell-area volunteers who can, in turn, develop the type of expert local knowledge that has formed the center's niche among wild horse organizations. Although outside groups contribute to a wide network of support, their visits to the range are, of necessity, limited in number and by season.

The Future of the PMWHR

The PMWHR has always enjoyed a special status as a herd management area, because it was the first public wild horse range, predating the 1971 Wild Free Roaming Horses and Burros Act; because the Pryor wild herd has a unique Spanish heritage; and because Lovell advocates have been consistently engaged with the Billings Field Office in implementing leading-edge wild horse management practices. On the other hand, the PMWHR has attracted considerable public attention because of Cloud, the palomino roan stallion who is the star of several PBS *Nature* documentaries. In part because of this publicity and a trend toward greater public awareness of the wild horse issue in general, nearly every BLM removal on the PMWHR has been subject to legal challenges from both outside organizations and a few individuals who live in Lovell and whose distrust of the BLM dates back to the early 1960s.

The Billings Field Office as well as the center have been lightning rods because so many people have watched the Cloud documentaries and follow the Pryor wild horses on a number of social media sites. The PMWHR is a beautiful and accessible place, compared to the vast expanses of many other herd management areas, particularly in Wyoming and Nevada. Visitors who are willing and able to tackle the rugged two-hour drive to the top of East Pryor Mountain during the summer months are rewarded by the presence of family bands, including foals, as well as bachelor stallion groups. There are very few places where it is possible to view wild horses displaying social behaviors in a natural setting, such as band stallions "snaking," or pushing their families away from other stallions; gently play fighting with their yearling and two-year-old colts; and tolerating the antics of their foals.

The question about whether the wild horses of the PMWHR are still of Spanish type hangs in the air. By adjusting the removal criteria within the parameters of the 2009 *HMAP* to emphasize continued representation of all existing bloodlines, the Billings Field Office has kept a genetic status quo. There is genetic diversity, but, as the Billings Field Office has pointed out, the herd has always had a mixed ancestry that includes Spanish blood. "The mix of breeds and historically introduced horses is directly responsible for the high level of genetic variation."[15]

It is now acknowledged by the BLM, as well as the center and many outside activist groups, even if it is not stated explicitly, that trying to breed back to the iconic Pryor Mountain Mustang by removing horses descended from outside introduced horses is now virtually impossible because of the interbreeding that has occurred within the herd.

Still, the issue of the herd's Colonial Spanish origins continues to surface as a policy alternative guiding management of the PMWHR. On March 22, 2013, the Billings Field Office issued a Draft Resource Management Plan and Environmental Impact Statement for the planning area that includes the PMWHR.[16] All PMWHR management documents—the 2009 *HMAP*, 2009 and 2012 removal plans, and 2011 fertility-control plan—are "tiered" to the current Resource Management Plan that was last approved in 1984. That plan covers multiple resource management objectives, including grazing allocations, wildlife improvements, wilderness study, and wild horse monitoring and range improvement. The Draft Resource Management Plan includes four alternatives for the way that the BLM manages the Pryor wild herd. These options include a combination of changes to the PMWHR boundary and criteria for managing the herd. The comments submitted will be one indicator of how the public perceives the Pryor wild herd.

One alternative would maintain the status quo. The second would close lands leased from private owners (for example, the Krueger property at the top of East Pryor Mountain) and return boundary lines to the 1968 Secretarial Order. The second alternative also reflects preferences against any management of the herd structure. It would allow the makeup of the herd to evolve through natural selection, although the law still requires the BLM to manage for an Appropriate Management Level. The third alternative would open the administrative pastures at the base of East Pryor Mountain, known as Britton Springs, to wild horse use and would open buffer areas adjacent to private lands. This alternative would manage for the "quintessential Pryor horse that is Dun or Grulla with striping and line back markings."[17] The fourth alternative, favored by the BLM, would keep the buffer areas closed, open the Britton Springs pastures to the PMWHR, and continue to manage for representation of all bloodlines and colors.[18]

Inasmuch as interbreeding has already influenced bloodlines, ensuring that all stallion-mare combinations are preserved for future generations is likely the most feasible option for preserving the genetic inheritance of the Colonial Spanish horse. Nevertheless, considering the importance attached to the Pryor Mountain Mustang and its Spanish heritage in advocating for the creation of the PMWHR, it is important to continue monitoring the genealogies of each Pryor wild horse. Public support for preserving the distinctive Spanish lineage of these horses appears to be declining, as indicated by recent comments in response to the 2012 bait-trap removal. Many comments reflected greater concern about inbreeding if the BLM were to manage for Spanish phenotype alone. Other reasons for favoring the current approach of representing all bloodlines and colors regardless of lineage likely motivated these comments, especially fears that individual favorites might be removed in the next removal. Several experts have commented (independently of each other) that many people will "love the Pryor horse to death."

Building Networks for the Future

This narrative is essentially a history about a community-based grassroots effort led by a group of men and women dedicated to the preservation of the Pryor Mountain Mustang. The Pryor Mountain Wild Horse Association, renamed the Pryor Mountain Mustang Association and then the Pryor Mountain Wild Mustang Center, established itself as having a distinctive local niche. Maintaining their organizational autonomy was, in part, a reaction against criticism of their preference for consultation over confrontation with the BLM. It was a successful strategy aimed at establishing a local presence—a "place at the table." This group of Lovell advocates has taken a consistent position over the years on several controversial issues: maintaining an Appropriate Management Level, using temporary fertility control, and conducting low-impact, small-scale removals.

Although the focus has slowly evolved from an exclusive focus on Spanish phenotype to a broader approach aimed at preserving all bloodlines represented on the range, the idea of selecting for *any* herd characteristics is in conflict with the views of many activists. Those who oppose selecting for a type of horse argue that it is a form of management that smacks of too much human control. They would prefer to see herd characteristics determined through natural selection, letting the fittest survive. The center continues to take positions that are at odds with most wild horse activists. As is the case with virtually all nonprofit organizations, however, the center will find it increasingly difficult to operate without a network of resources outside, as well as within, Lovell. In the past the center has insisted on being geographically and politically isolated; however, limited human and financial resources may be a catalyst for change.

It remains to be seen if the center can build a network without jeopardizing its identity. Moreover, the local culture of Lovell is skeptical of radical politics and dismissive of "horse huggers." Thus, the challenge the center faces is to develop a network of support within Lovell based on residents' identification with the mustang "brand" and the values that are unique to their community, while at the same time reaching out to individuals and groups who visit the PMWHR on a regular basis. One example is Sandy Elmore, who lives in Montana and makes frequent trips to the range, posting photographs and updates on her *Wild in the Pryors* blog.

The center could benefit from her frequent updates, while maintaining a Facebook page, blog, or newsletter with its own comments and reflections. Likewise, photographers who conduct workshops or visit the area on a regular basis can contribute to center updates of its wild horse list by documenting newborn foals and other changes in the makeup of family bands.[19] The center could remain responsible for its horse list and "fact-check" the information provided by photographers, including color, face, and leg markings of individual horses and their movement between family bands or bachelor groups. The BLM's longstanding trust in the center's database depends on the validity and reliability of the center's data, and this relationship is at least as important as—if not more significant than—the network of support needed to sustain the center in the future.

This tension between the need to build networks of internal and external support and fulfill the responsibilities laid out in the 2008 Memorandum of Understanding with the Billings Field Office presents a major challenge for whoever assumes leadership of the center in the near future. The Pryor wild horses are the subject of great concern to outside organizations, such as the Cloud Foundation, whose director, Ginger Kathrens, and TCF staff members also visit the PMWHR from Colorado Springs on a regular basis. TCF is a member of the American Wild Horse Preservation Campaign, a national coalition of fifty-eight wild horse organizations with a much broader policy agenda.[20] The major challenge the center faces will be to clearly define its mission, identify the specific kinds of support it needs, and decide how to manage exchanges of support from, and contributions to, members of the networks it may choose to develop in the future.

Appendix

Methodology

This book is a case study of a Wyoming community in which a cadre of wild horse advocates worked to save, then preserve, a local herd of wild horses. The community of Lovell is unique because it is the only known place where citizens sustained such an effort over a long period of time—in this instance nearly fifty years. The wild herd itself is unique because it is one of the last wild herds carrying genetic markers for the original Colonial Spanish horse, the battle mount of the conquistadores. Finally, the working relationship between Lovell advocates and the Billings, Montana, field office of the Bureau of Land Management has been positive and constructive. One does not normally see the words *consultation* and *wild horses* used in the same sentence. For all these reasons, the focus of this book is on a single community, one wild horse herd, and a field office of a major federal agency.

The methods used for this qualitative research project were twofold: document review and field interviews with key informants. The vast majority of documents were materials in the Pryor Mountain Wild Mustang Center Archives in Lovell, Wyoming. The Billings Field Office also provided administrative records dating from 1968 to 2009. Key informants received the following information: a synopsis of the book project, a letter of invitation approved by the IRB (Institutional Review Board) at the University of Nebraska–Omaha, and a summary of the open-ended questions. In order to ensure ethical access to key informants, the author sent this correspondence through the Pryor

Mountain Wild Mustang Center, unless contact information was publicly available. In no cases did the author make "cold calls."

IRB# 087-12-EX

Dear

I am writing a book on the Pryor Mountain Wild Horse Range with a special focus on the relationship between local residents and the Pryor Mountain Mustangs. I am working on this project under a conditional contract with University of Nevada Press. My goal is to complete research for my book during June and July, 2013.

I am writing to ask if you would be willing to spend an hour or so with me while I am in Lovell this summer, exploring the questions in the attached field study guide about which you have personal knowledge or experience. I have approval from the Human Subjects board at my university based on my guarantee of privacy and confidentiality.

I promise not to reveal the individual identity or to publish the results of an interview in any way that would reveal the identity of those I have interviewed. I will, however, publish a general thank you to all who contributed to my project at the end of the book, e.g. "past and present members of the Pryor Mountain Wild Mustang Center, Pryor Mountain Mustang breeders," etc.

If you are willing and available to meet with me, I arrive in Lovell, on June 4, 2012 (and June 15, 2013) and will be living there until the end of July. I would appreciate a reply by either email to: creed@unomaha.edu or to my cell # (402) 672-9227 either before or during the month of June. I look forward to hearing back from you.

Sincerely,

Christine Reed
creed@unomaha.edu
School of Public Administration
University of Nebraska at Omaha
Omaha, NE 68132
402-554-2864 (office)
402-672-9227 (mobile)

Field Study Guide

1. What can you tell me about the formation of the Pryor Mountain Wild Horse Association? Who were the founding officers and who were the first members? Why did they found the Wild Horse Association? What was the relationship between the WH Association and the Billings MT field office of the BLM when the Association was founded?
2. What can you tell me about the formation of the Pryor Mountain Mustang Breeder Association? Who founded the association and what were the major goals of the organization? What can you tell me about the foundation stock of the registry and the criteria for registration? How many Pryor Mustangs have been registered since the start of the Breeder Association? How many registered PMM's were bred by association members versus adopted from the Pryor Mountain Wild Horse Range? How many owners are still active in the Breeder Association? What has been the relationship between the Breeder Association and the Billings MT field office of the BLM over the years?
3. What can you tell me about the formation of the Pryor Mountain Wild Mustang Center? Who founded the Center and who served on the original board? Who has served on the board since its founding? Who are the current board members? What was the original mission of the Center and how has it changed over the years? What were its primary activities or projects at the beginning and how have they changed over the years? What has been the relationship between the Center and the Billings, MT field office of the BLM over the years?
4. What can you tell me about the involvement of Lovell citizens with the Pryor Mountain Wild Horse Range over the years? How have Lovell citizens been involved? Have they adopted Pryor Mountain Mustangs during periodic gathers by the BLM? What about the Breeder Association: have Lovell citizens purchased PMM's from local breeders? What was the origin of Lovell "Mustang Days"? How has "Mustang Day" activities supported the Pryor Mountain Mustangs? How would you characterize the identification of Lovell citizens or "alumni" of Lovell with the Pryor Mountain Mustangs?

5. What is the current status of the Breeder Association? What are the official plans for BLM management of the Pryor Mountain Wild Horse Range? What is the Center's position on the future of both domestic and wild Pryor Mustangs?

Notes

One | Trespass Horses

1. Tom Mast, "Pryor Mustangs Embody New World History," *Casper Star-Tribune*, February 2, 1997.

2. See Patricia M. Fazio, "The Fight to Save a Memory," *Annals of Wyoming: The Wyoming History Journal* (Spring 1997): 29, for an extended discussion of various theories about the first horses that formed the foundation stock of the Pryor wild horse herd.

3. "Establishment of Pryor Mountain Wild Horse Range," 33 *Fed. Reg.* 173 (September 12, 1968).

4. See J. Edward Desteiguer, *Wild Horses of the West* (Tucson: University of Arizona Press, 2011), for a comprehensive history of wild horses on the North American continent.

5. Joan Brownell, "Horse Distribution in the Pryor Mountain Region Preceding Creation of the PMWHR," Bureau of Land Management: Administrative Record for Environmental Assessment MT-C010-2009-35 (1999), 9.

6. D. Phillip Sponenberg, "North American Colonial Spanish Horse Update," http://www.frontiernet.net/~RanchoTamarisque/Sponenberg2003update.

7. Brownell, "Horse Distribution in the Pryor Mountain Region," 21.

8. Charles O. Williamson to Norma Bearcroft, July 12, 1967, PMWMC Archives.

9. Hope Ryden to Linda Coates-Markel, January 19, 2000, PMWMC Archives.

10. Kurt McNabb, *Lloyd Tillett: The Man Who Saved America's Wild Horses* (1974), audio CD.

11. Fazio, "Fight to Save a Memory," 32.

12. Confidential interview, June 14, 2012.

13. Fazio, "Fight to Save a Memory," 33.

14. Robert Glicksman, *Modern Public Land Law* (St. Paul: West, 2006).

15. Fazio, "Fight to Save a Memory," 35.

16. Beverly L. Robertson, "Crooked Creek Wild Horses Get Reprieve from BLM," *Lovell Chronicle,* April 14, 1966.

17. Confidential interview, July 10, 2013.

18. Floyd Schwieger to Representative Teno Roncalio, n.d., PMWMC Archives.

19. "Wild Horses Arouse Interest," *Lovell Chronicle,* March 31, 1966.

20. Don Hegg, "Horse Herd Not Wild Insists BLM," *Billings Gazette,* n.d.

21. Fazio, "Fight to Save a Memory," 40.

22. Ibid., 41.

23. BLM associate director John Crow to Senator Mike Mansfield, April 5, 1968, PMWMC Archives.

24. Ibid.

25. "McGee Wants Land for Wild Horse Herd," *Billings Gazette,* April 5, 1968.

26. The proposed road would have cut through the Crow Reservation and was never completed. It now dead-ends at Barry's Landing.

27. Fazio, "Fight to Save a Memory."

28. Confidential interviews, June 19 and July 10, 2013. "Taking" refers to a clause in the Fifth Amendment to the Constitution, prohibiting the federal government from taking property by eminent domain without just compensation to the owners.

29. Confidential interview, June 19, 2013.

30. PMWMC permanent display.

31. Letters addressed to Governor Hansen and to the Lovell Chamber of Commerce, PMWMC Archives.

32. Alan S. Kania, *Wild Horse Annie: Velma Johnston and Her Fight to Save the Mustang* (Reno: University of Nevada Press, 2013).

33. The number of twenty head likely referred to the number permitted on the Crooked Creek grazing allotment.

34. Velma B. Johnston to Mansfield, April 9, 1968, PMWMC Archives.

35. Floyd Schwieger to Mansfield, April 11, 1968, PMWMC Archives.

36. Johnston to Schwieger, April 24, 1968, PMWMC Archives.

37. "Lovell Association Formed to Sponsor Wild Horse Herd," *Lovell Chronicle,* June 20, 1968.

38. "He's Out to Save Horses," *Billings Gazette,* June 28, 1968.

39. Hope Ryden, *America's Last Wild Horses* (Guilford: Lyons Press, 1999).

40. The photograph collection of Hope Ryden is located at the PMWMC Archives.

41. Fazio, "Fight to Save a Memory," 45.

42. Lloyd Tillett to Hope Ryden, August 5, 1968, PMWMC Archives.

43. Charles Wagner to Ryden, August 1, 1968, PMWMC Archives.

44. Ibid.

45. Hope Ryden (personal notes), August 12, 1968, PMWMC Archives.

46. Johnston to Schwieger, August 18, 1968, PMWMC Archives.

47. David Dominick to Boyd Rasmussen, August 21, 1968, PMWMC Archives.

48. Fazio, "Fight to Save a Memory," 46.

49. "Suit Filed to Save Mustang Horse Herd," *Casper Star-Tribune*, August 29, 1968.

50. "BLM Establishes Wild Horse Range in Montana," *Lovell Chronicle*, September 12, 1968.

51. Declaration of Hope Ryden in *Cloud Foundation, Inc., Front Range Equine Rescue, et al., v. Dirk Kempthorne*, September 8, 2006.

52. "Establishment of the Pryor Mountain Wild Horse Range."

53. *Congressional Record*, Senate, September 12, 1968, 26658.

54. "Establishment of the Pryor Mountain Wild Horse Range."

55. *Congressional Record*, Senate, September 12, 1968, 26658.

56. "Advisory Group Hears Local Views on Wild Horse Herd," *Lovell Chronicle*, October 24, 1968.

57. This fence became controversial when the BLM announced its intentions to remove all or most of the wild herd, but its original purpose was a wild horse refuge sponsored by a local group, such as the Pryor Mountain Wild Horse Association.

58. McNabb, *Lloyd Tillett*.

59. Memo from wild horse wrangler to area manager, Billings Resource Area, July 27, 1990, PMWMC Archives.

60. Ross Carletta, "Wild Nags to Be Weeded," *Billings Gazette*, March 27, 1969.

61. "Wild Horse Committee Meets Here with BLM Officials," *Lovell Chronicle*, May 28, 1970.

62. Jean Wagner, "BLM Plans Mustang Reduction," *Lovell Chronicle*, October 7, 1971.

63. "Horse Culling Plans Premature," *Lovell Chronicle*, October 7, 1971.

64. "Trap Wild Horses," *Lovell Chronicle*, November 4, 1971.

65. Ibid.

66. This arrangement predated the adoption clause in the 1978 Public Rangeland Improvement Act, which amended the 1971 Wild Free Roaming Horses and Burros Act.

67. "Thanks Lovell—WHOA," *Lovell Chronicle*, March 16, 1972.

68. Edwin Baidliz to Schwieger, May 2, 1978, PMWMC Archives.

Two ⁝ Growing Pains

1. US Bureau of Land Management, *Pryor Mountain Wild Horse Range/Territory Environmental Assessment and Herd Management Area Plan* (MT-010-08-24 2009), 3.

2. "Establishment of Pryor Mountain Wild Horse Range" (see chap. 1, n. 3).

3. Public Law 92-195, December 17, 1971, sec. 1 (emphasis added).

4. Ron Hall, *Wild Horse Biology and Alternatives for Management: Pryor Mountain Wild Horse Range* (Bureau of Land Management Billings District, 1972), BLM-002284–87.

5. US Bureau of Land Management Billings District and USFS Beartooth Ranger District, *Pryor Mountain Complex Land Use Decisions* (May 1974). A county road separates the boundary fence encompassing the remaining portion of the Crooked Creek allotment from the Tillett property line.

6. US Bureau of Land Management, *Herd Management Area Plan, Pryor Mountain Wild Horse Range* (Miles City District Office, June 1984), 1.

7. Hall, *Wild Horse Biology*, BLM-002258.

8. Ibid., BLM-002282.

9. http://www.albc-usa.org.

10. BLM, *Herd Management Area Plan* (1984), 11–12.

11. Ibid., 15.

12. Jay F. Kirkpatrick, *Into the Wind* (Minocqua, WI: NorthWord Press, 1994), 12.

13. "Do Pryor Wild Horses Have Roots in Spain?," *Billings Gazette*, March 23, 1983.

14. Ibid.

15. E. Gus Cothran, "Genetic Analysis of the Pryor Mountain Wild Horse Reserve" (unpublished manuscript, 1992).

16. "Bureau Adds Five Horses to Pryor's Wild Herd," *Billings Gazette*, November 16, 1989.

17. BLM, *Herd Management Area Plan* (1984), 25.

18. Jerome Jack to Russell Gaspar, 1982, PMWMC Archives.

19. There were a series of removals starting in 1982 (forty-three), 1983 (twenty-one), 1984 (thirteen), 1985 (twenty-five), 1987 (twenty-three), 1988 (twenty-six), and 1989 (twenty-one), and BLM records show that all removed horses were adopted.

20. Ryden to Schwieger, June 6, 1990, PMWMC Archives.

21. Memo, "The Pryor Mountain Wild Horse Association," November 16, 1989, PMWMC Archives.

22. Ibid.

23. Memo, "Horse Management Section of the Pryor Mountain Wild Horse Association," November 16, 1989, PMWMC Archives.

24. "First Fall Meeting a Success," *Lovell Chronicle,* September 13, 1990.

25. E. Gus Cothran to Schwieger, January 19, 1990, PMWMC Archives.

26. "First Fall Meeting a Success."

27. Memo, "Wild Horse Coalition Meeting," January 10, 1991, PMWMC Archives.

28. "Horse Roundup Questioned," *Lovell Chronicle,* March 21, 1991.

29. "Roundup Proceeding Well," *Lovell Chronicle,* April 4, 1991.

30. "Wild Horse Herd Must Be Kept at 121," *Lovell Chronicle,* April 18, 1991.

31. "Work Together for the Good of the Horses," *Lovell Chronicle,* March 28, 1991.

32. "Herd Plan Changes Fought by Public," *Lovell Chronicle,* October 31, 1991.

33. "Memo Directive to the Horse and Range Committee from the Board of Directors of PMMA Regarding Assistance to BLM in Culling Horses from the Herd," October 1, 1992, PMWMC Archives.

34. Minutes of the PMMBA, March 1992, PMWMC Archives.

35. PMMBA bylaws to constitution, 1992, PMWMC Archives.

36. D. P. Sponenberg, "Evaluation of Pryor Mountain Herd Area BLM Horses" (unpublished manuscript, August 1993).

37. Ibid.

38. D. P. Sponenberg, "Comments on Pryor Horse Management" (unpublished manuscript, April 1994).

39. D. P. Sponenberg, "Report on the Pryor Mountain Mustangs" (unpublished manuscript, October 10, 1994).

40. Confidential interview, June 26, 2012.

41. "Pryor Mountain Mustang Association Evaluation of 1994 Roundup," n.d., PMWMC Archives.

42. Memo, "BLM Response to Pryor Mountain Wild Horse Population Report," January 1995, PMWMC Archives.

43. Memo, "Flowchart for the Pryor Mountain Mustang Association (PMMA) Meeting," February 27, 1997, PMWMC Archives.

44. Minutes of the PMMA, September 11, 1997, PMWMC Archives.

45. Memo, "Revised Super Horse List," fall 1997, PMWMC Archives.

Three Barn Raising

1. McNabb, *Lloyd Tillett* (see chap. 1, n. 10).

2. PMMBA *Newsletter,* April 2007.

3. PMWMC, "Articles of Incorporation of Pryor Mountain Wild Horse Center," October 27, 1998, PMWMC Archives.

4. PMWMC, "Bylaws for Pryor Mountain Wild Horse Center, Inc.," March 5, 1999, PMWMC Archives.

5. PMWMC, "Draft Promotional Brochure for the Pryor Mountain Wild Horse Center," PMWMC Archives.

6. Memo, "President's Report to the Pryor Mountain Wild Horse Center Board," August 5, 1999, PMWMC Archives.

7. Minutes of the PMWMC meeting, November 18–19, 1999, PMWMC Archives.

8. PMWMC *Newsletter,* fall 2002.

9. Many other individuals who served on the board or attended meetings as interested members deserve to be mentioned. The following list may exclude some individuals, and the author apologizes for any oversights: Dale Hartman, Philip Schaelling, Bill Roberts, Earl Lukkes, Wes Meeker, Dick Doerr, Todd Wacaser, Jerry Tippetts, Jim Minchow, La Moine Sorenson, and Bill and Kathy Johnsey.

10. Minutes of the PMWMC board meeting, May 12, 2000, PMWMC Archives.

11. Ibid., January 11, 2001.

12. David Peck, "Wild Mustang Center Has Great Potential," *Lovell Chronicle,* May 24, 2001.

13. PMWMC *Newsletter,* fall 2002.

14. PMWMC *Newsletter,* spring 2003.

15. Minutes of the PMWMC meeting, February 8, 2005, PMWMC Archives.

16. PMWMC *Newsletter,* fall 2005.

17. Ibid.

18. Tom Mast, "Pryor Mustangs Embody New World History," *Casper Star-Tribune,* February 2, 1997.

19. PMWMC *Newsletter,* fall 2005.

20. The individuals who donated their time included Keith Grant, Kim Newlin, Mike Leonhardt, Jay Bischoff, Dale Hartman, Bill Johnsey, Cliff Revelle, Lomand Fowler, George Pries, Earl Lukkes, William Bridges, Ed Roller, and John Nickle.

21. Minutes of the PMWMC board meeting, May 11, 2005, PMWMC Archives.

22. PMWMC *Newsletter,* fall 2006.

23. PMWMC *Newsletter,* fall 2007.

24. Pamphlet of the PMWMC, *Meet the Mustangs,* July 17, 2004, PMWMC Archives.

25. PMWMC *Newsletter,* fall 2005.

26. Hall, *Wild Horse Biology* (see chap. 2, n. 4).

27. Schwieger to Sandy Brooks, January 10, 2000, PMWMC Archives.

28. Schwieger to Ann Venamen, January 25, 2001, PMWMC Archives.

29. Nickle to Venamen, February 2, 2001, PMWMC Archives.

30. *Cloud Foundation v. Kempthorne,* 546 F.Supp.2d 1003 (D. Mont. 2008).

31. Minutes of the PMWMC board meeting, March 11, 2006, PMWMC Archives.

32. The plant was identified earlier as *Shoshonea pulvinata.* Billings Resource Area Office, "Summary of the Pryor Mountain Wild Horse Range Meeting," January 10, 1991, PMWMC Archives.

33. Update of the PMWMC board president, John Nickle, February 15, 2008, PMWMC Archives.

34. Report of the PMWMC meeting, February 22, 2008, PMWMC Archives.

35. Ibid.

36. Report of the PMWMC meeting, February 27, 2008, PMWMC Archives.

37. PMWMC *Newsletter,* fall 2006.

38. Ibid.

39. PMWMC *Newsletter,* fall 2005.

40. Interview with Matthew Dillon, January 28, 2010.

41. Bureau of Land Management, "Memorandum of Understanding Between United States Department of Interior Bureau of Land Management Billings Field Office and Pryor Mountain Wild Mustang Center," BLM-MOU-MT010-2007-01.

42. According to an interview with a center board member, in 2004 a mountain lion and her two young cubs killed all but one foal, but it is unlikely that predation would have led to the deaths of the entire foal crop, especially since deaths occurred throughout the area, from the top of East Pryor Mountain to the lower elevations in the Dryhead.

43. Minutes of the PMWMC board meeting, May 15, 2009, PMWMC Archives.

44. Bureau of Land Management, *Pryor Mountain Wild Horse Range/ Territory Environmental Assessment* (MT-010-08-24) and *Herd Management Area Plan* (May 2009).

Four ⁞ A Perfect Storm

1. BLM, *Herd Management Area Plan* (2009), 27.

2. BLM, *Herd Management Area Plan* (1984), 25.

3. Lori S. Eggert et al., "Pedigrees and the Study of the Wild Horse Population of Assateague Island National Seashore," *Journal of Wildlife Management* 74 (2010): 963–73.

4. Interior Board of Land Appeals, 2009-246, November 3, 2009.

5. Ibid.

6. Ibid., January 15, 2010.

7. Ibid.

8. Bureau of Land Management Billings Field Office, *Pryor Mountain Wild Horse Range 2009 Gather Plan and Environmental Assessment* DOI-BLM-MT-CO10-2009-35-EA, 5.

9. Ibid., 10.

10. Nickle and Matt to Jim Sparks, Billings Field Office, August 18, 2009, *Administrative Record for Environmental Assessment* MT-CO10-2009-35.

11. Analysis by the author.

12. Dr. Phil Sponenberg, "Comments on Pryor Mustang Herd, June 2009," in *Administrative Record for Environmental Assessment* MT-CO10-2009-35.

13. PMWMC *Newsletter*, winter 2009.

14. Draft letter of the Pryor Mountain Wild Mustang Center and Friends of a Legacy to US Bureau of Land Management, Washington, DC, n.d., PMWMC Archives.

15. Senate 1579 amending Public Law 92-195, Wild Free Roaming Horses and Burros Act of 1971.

16. PMWMC, *Newsletter*, summer 2010.

17. Bureau of Land Management Billings Field Office, *Pryor Mountain Wild Horse Range Fertility Control Environmental Assessment December* 2010 DOI-BLM-MT-0010-2011-0004-EA.

18. Ibid., 8.

19. Ibid., 12.

20. Author's observations at the PMWMC during June and July 2011.

21. PMWMC *Newsletter*, summer 2010.

22. Author's observations at the center during the summer of 2012.

23. Fran Lynghaug, *Official Horse Breeds Standards Guide* (Minneapolis: Voyageur, 2009), 106.

24. Minutes of the PMWMC board meeting, November 19, 2009, PMWMC Archives.

25. http://www.arrowheadmustangs.com.

26. Confidential interviews, June 2012 and January 2013.

27. Communication with Dr. Phil Sponenberg, September 12, 2012.

28. Bureau of Land Management Billings Field Office, *Pryor Mountain Wild Horse Range Non-helicopter Gather Environmental Assessment* 2012 DOI-BLM-MT-0010-2-12-0001-EA, 4.

29. Ibid., 5.

30. Ibid., 6.

31. Ibid., 10.

32. Ibid., 9–10.

33. National Park Service personnel assisted with the bait-trap and removal operations.

34. There was a brief delay in June because of an unsuccessful petition in the Billings federal district court for a temporary restraining order filed by three plaintiffs, Cathy Bryarly, Jerri Tillett, and Nikki Spotted Owl. See http://www.wildhoofbeats.com, July 28, 2012.

35. Brett French, "Trapping Mountain Wild Horses No Quick Task," *Billings Gazette*, August 2, 2012.

36. http://www.blm.gov/mt/st/en.html.

37. *Pryor Wild* blog, http://www.pryormustangs.org, August 5, 2012.

38. See as an example http://wildinthepryors.com, October 2012.

39. US Bureau of Land Management, "BLM Acting Director Mike Pool Presents Third Annual Director's Awards to Employees for Extraordinary Public Service," October 25, 2012.

Five | Looking to the Future

1. Visit by the author to Arrowhead Ranch and conversation with Red Tolman on October 13, 2012.

2. This conclusion is based on the author's research on the number of registrations during the past fifteen years.

3. Communication with Cedric Berthouzoz, October 7, 2012.

4. Ibid., June 20, 2013.

5. Ibid., October 7, 2012.

6. Communication with Florence Mouninou, February 12, 2013.

7. Ibid., February 14, 2013.

8. http://thecloudfoundation.org/about-us/meet-the-freedom-fund-horses.

9. Interview with Steve Smith, July 13, 2013.

10. http://www.albc-usa.org.

11. The Sweet Medicine Wild Horse Sanctuary in Fishtail, Montana, also started with horses adopted from the PMWHR, but the owner, center board member Jim Edwards, has not been a member of the breeders association.

12. BLM, "Memorandum of Understanding" (see chap. 3, n. 41).

13. Lori Graham, the center director who succeeded Matt Dillon, resigned in the fall of 2012. Aubrey Jolley, the center board secretary-treasurer, assumed the position in November 2012, but resigned in June 2013. John Nickle is now serving as interim director.

14. Interview with Sue Taylor, June 18, 2013.

15. BLM Billings Field Office, *Fertility Control* (2010) 15 (see chap. 4, n. 17).

16. BLM Billings Field Office, *Billings and Pompeys Pillar National Monument Resource Management Plan Revision Fact Sheet*, July 1, 2013.

17. Ibid.

18. Ibid.

19. Professional wild horse photographer Lynne Pomeranz conducts annual workshops on the PMWHR. Another photographer who makes regular visits to the range is Shawn Ivie, who has his own website and blog under the domain name "Wyoman."

20. http://www.wildhorsepreservation.org.

Selected Bibliography

Desteiguer, J. Edward. *Wild Horses of the American West.* Tucson: University of Arizona Press, 2011.

Kania, Alan S. *Wild Horse Annie: Velma Johnston and Her Fight to Save the Mustang.* Reno: University of Nevada Press, 2013.

Kirkpatrick, Jay F. *Into the Wind.* Minocqua, WI: NorthWord Press, 1994.

Lynghaug, Fran. *Official Horse Breeds Standards Guide.* Minneapolis: Voyageur, 2009.

Ryden, Hope. *America's Last Wild Horses.* Guilford, CT: Lyons, 1999.

Sponenberg, D. Phillip, and Donald E. Bixby. *Managing Breeds for a Secure Future: Strategies for Breeders and Breed Associations.* Pittsboro, NC: American Livestock Breeds Conservancy, 2007.

Index